STUDY GUIDE
AND
SELECTED SOLUTIONS MANUAL

Charles H. Corwin

American River College

Introductory Chemistry

CONCEPTS AND CRITICAL THINKING
SIXTH EDITION

Prentice Hall

Boston   Columbus   Indianapolis   New York   San Francisco   Upper Saddle River
Amsterdam   Cape Town   Dubai   London   Madrid   Milan   Munich   Paris   Montréal   Toronto
Delhi   Mexico City   São Paulo   Sydney   Hong Kong   Seoul   Singapore   Taipei   Tokyo

Acquisitions Editor: Terry Haugen
Editor in Chief, Chemistry and Geosciences: Nicole Folchetti
Marketing Manager: Erin Gardner
Assistant Editors: Laurie Hoffman and Carol DuPont
Managing Editor, Chemistry and Geosciences: Gina M. Cheselka
Project Manager: Wendy A. Perez
Operations Specialist: Maura Zaldivar
Supplement Cover Designer: Paul Gourhan
Cover Photo Credit: Kazuyoshi Nomachi/Corbis

Printed in the United States of America

10 9 8 7 6 5 4 3 2

ISBN-13:   978-0-321-67514-9
ISBN-10:   0-321-67514-2

Prentice Hall
is an imprint of

PEARSON
www.pearsonhighered.com

Contents

Study Guide
with Self-Test Questions and Crossword Puzzles of Key Terms

Selected Solutions Manual
for Odd-Numbered Textbook Exercises

Flashcards
Common Monoatomic and Polyatomic Cations and Anions

Preface

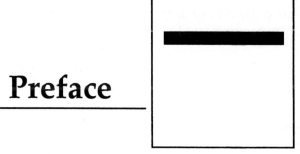

To the Student

I sincerely hope that this *Study Guide & Solutions Manual* will enable you to study chemistry more effectively while using the textbook **Introductory Chemistry, Sixth Edition**. This *Study Guide & Solutions Manual* is designed so that you can test your understanding of each topic and discover areas that require additional study. Thus, you will find self-test questions for each objective in the text.

There is also a crossword puzzle following the self-test questions to assist you in learning the key terms in each chapter. The answers to the self-test questions, as well as the crossword puzzle, are included at the end of each set of chapter questions of this *Study Guide & Solutions Manual*.

The following guidelines will help you achieve maximum success in your introductory chemistry course.

- **First**, take careful notes in lecture and pay close attention to the assignments given by your instructor.

- **Second**, read the assigned sections in the text and study the example exercises. Do the end-of-chapter exercises and check your answers in *Appendix I*. If you cannot solve a problem or have difficulty with an exercise, refer to the complete solution at the end of this study guide.

- **Third**, answer the self-test questions in this study guide to diagnose any lack of understanding in the assignments by your instructor. If you answer a question incorrectly, refer to the corresponding section in the textbook.

- **Fourth**, take the practice Self-Test at the end of each assigned chapter in the textbook and check your answers in *Appendix J*. If you answer a question incorrectly, refer to the corresponding section in the textbook.

If you observe the above suggestions, this *Study Guide & Solutions Manual* will make your job of learning chemistry much more efficient. For those topics that still seem unclear, visit your instructor during office hours. Most instructors enjoy helping students, especially those who are making a serious effort to do the assignments and learn the material.

If you want to share your chemistry experience, please send me an e-mail or drop me a note at the address below. I would like to hear about your success as well as any suggestions for improving these study aids.

Charles H. Corwin
Department of Chemistry
American River College
Sacramento, CA 95841
corwinc@arc.losrios.edu

Study Guide

with

Self-Test Questions

and

Crossword Puzzles of Key Terms

Introduction to Chemistry

Section 1.1 *Evolution of Chemistry*

1. What four elements composed everything in nature according to the beliefs of the ancient Greeks?
    - (a) air, earth, salt, and water
    - (b) air, ashes, fire, and water
    - (c) air, earth, fire, and water
    - (d) smoke, earth, fire, and water
    - (e) none of the above

2. Which of the following is a basic step in the scientific method?
    - (a) perform an experiment and collect data
    - (b) analyze experimental data and propose a hypothesis
    - (c) test a hypothesis and state a theory or law
    - (d) all of the above
    - (e) none of the above

Section 1.2 *Modern Chemistry*

3. In which of the following industries does chemistry play an important role?
    - (a) agriculture
    - (b) electronics
    - (c) paper
    - (d) transportation
    - (e) all of the above

4. Which of the following is *not* derived from a petrochemical?
    - (a) a gold coin
    - (b) a plastic toy
    - (c) an insecticide
    - (d) a synthetic fabric
    - (e) a blue dye

Section 1.3 *Learning Chemistry*

5. In a survey by the *American Chemical Society* of entering college students, which science course was considered the most relevant to their daily lives?
    (a) biology
    (b) chemistry
    (c) geology
    (d) physics
    (e) physiology

6. Which of the following is *not* a positive association with chemistry?
    (a) chemistry benefits society
    (b) chemistry has biomedical applications
    (c) chemistry involves dangerous experiments
    (d) chemistry topics are relevant to our daily lives
    (e) chemistry has many career opportunities

7. Connect the following nine dots using:

    (a) 5 straight, continuous lines.

    ```
    •   •   •

    •   •   •

    •   •   •
    ```

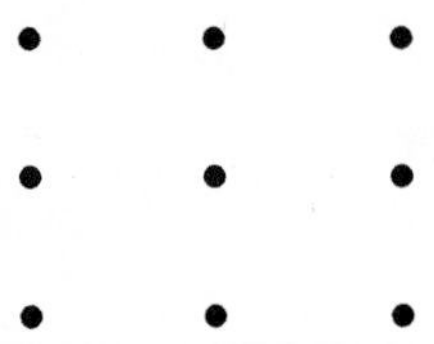

    (b) 4 straight, continuous lines.

    ```
    •   •   •

    •   •   •

    •   •   •
    ```

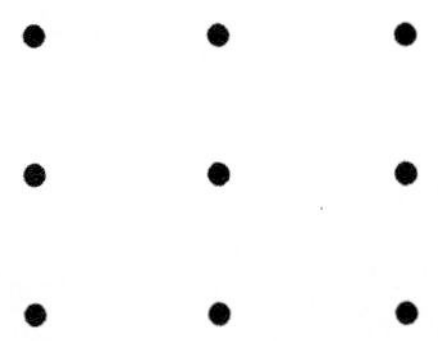

    (c) 3 straight, continuous lines.

    ```
    •   •   •

    •   •   •

    •   •   •
    ```

Chapter 1 Key Terms

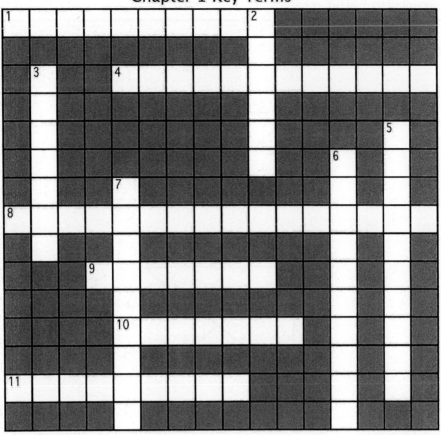

Across
1. a scientific procedure for collecting data
4. the study of biological substances and their reactions
8. a systematic investigation of nature (2 words)
9. the chemistry of substances containing carbon
10. a pseudoscience for changing lead into gold
11. the study of the composition of matter

Down
2. an extensively tested scientific proposal
3. the methodical exploration of nature
5. an initial explanation of experimental results
6. $P_1V_1 = P_2V_2$ (2 words)
7. substances not containing carbon

1. c 2. d 3. e 4. a 5. b 6. c

7. (a)

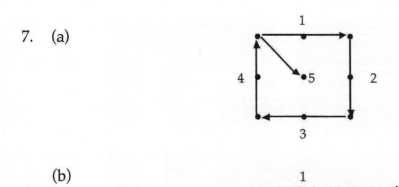

(b)

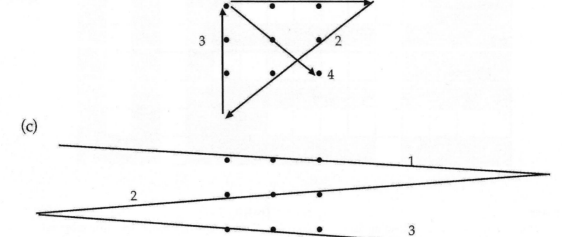

(c)

Chapter 1 Answers to Crossword Puzzle

Across
1. experiment
4. biochemistry
8. scientific method
9. organic
10. alchemy
11. chemistry

Down
2. theory
3. science
5. hypothesis
6. natural law
7. inorganic

Scientific Measurements

Section 2.1 *Uncertainty in Measurements*

1. Which of the following quantities can be measured exactly?
   - (a) length
   - (b) mass
   - (c) volume
   - (d) time
   - (e) no quantity can be measured exactly

2. Which of the following balances for measuring mass has the least uncertainty?
   - (a) analytical balance, ± 0.0001 g
   - (b) triple-beam balance, ± 0.01 g
   - (c) double-pan balance, ± 1 g
   - (d) electronic balance, ± 0.001 g
   - (e) platform balance, ± 0.1 g

Section 2.2 *Significant Digits*

3. How many significant digits are in the length measurement 20.05 meters?
   - (a) 1
   - (b) 2
   - (c) 3
   - (d) 4
   - (e) none of the above

4. How many significant digits are in the mass measurement 0.0550 kilogram?
   - (a) 2
   - (b) 3
   - (c) 4
   - (d) 5
   - (e) none of the above

Section 2.3 *Rounding Off Nonsignificant Digits*

5. Round off the following measurement to three significant digits: 107.500 g.
    - (a) 107 g
    - (b) 108 g
    - (c) 107.0 g
    - (d) 107.4 g
    - (e) 107.5 g

6. Round off the measurement 2345 cm to three significant digits.
    - (a) 230 cm
    - (b) 234 cm
    - (c) 235 cm
    - (d) 2340 cm
    - (e) 2350 cm

Section 2.4 *Adding and Subtracting Measurements*

7. Add 7.77 g to 51.665 g and round off the sum to the proper significant digits.
    - (a) 59.0 g
    - (b) 59.4 g
    - (c) 59.43 g
    - (d) 59.44 g
    - (e) 59.436 g

8. Subtract 2.25 cm from 11.5 cm and round off the difference to the proper significant digits.
    - (a) 9.0 cm
    - (b) 9.2 cm
    - (c) 9.20 cm
    - (d) 9.25 cm
    - (e) 9.3 cm

Section 2.5 *Multiplying and Dividing Measurements*

9. Multiply 2.505 m times 1.75 m and round off the product to the proper number of significant digits.
    - (a) 4.0 m^2
    - (b) 4.00 m^2
    - (c) 4.38 m^2
    - (d) 4.384 m^2
    - (e) 4.40 m^2

10. Divide 124.7 miles (mi) by 2.25 hours (h) and round off the quotient to the proper number of significant digits.
    (a) 50 mi/h
    (b) 55 mi/h
    (c) 55.0 mi/h
    (d) 55.4 mi/h
    (e) 55.5 mi/h

Section 2.6 *Exponential Numbers*

11. Express 0.000 000 000 000 000 000 000 000 000 000 010 as an exponential number.
    (a) 1.0×10^{-10}
    (b) 1.0×10^{-32}
    (c) 1.0×10^{-33}
    (d) 1.0×10^{32}
    (e) 1.0×10^{33}

12. Express the exponential number 5×10^{-9} as an ordinary number.
    (a) 0.000 09
    (b) 0.000 000 05
    (c) 0.000 000 005
    (d) 0.000 000 000 5
    (e) 5,000,000,000

Section 2.7 *Scientific Notation*

13. One gram of carbon contains 50,200,000,000,000,000,000,000 atoms. Express this number of carbon atoms in scientific notation.
    (a) 5.02×10^{21}
    (b) 5.02×10^{22}
    (c) 5.02×10^{23}
    (d) 50.2×10^{21}
    (e) 50.2×10^{22}

14. A carbon atom has a mass of 0.000 000 000 000 000 000 000 000 0199 g. Express this mass of a carbon atom in scientific notation.
    (a) 1.99×10^{-9} g
    (b) 1.99×10^{-21} g
    (c) 1.99×10^{-22} g
    (d) 1.99×10^{-23} g
    (e) 1.99×10^{-24} g

Section 2.8 *Unit Equations and Unit Factors*

15. Which of the following unit factors is derived from: 1 mile = 1.61 kilometer?
    (a)   1 mile/1 kilometer
    (b)   1.61 mile/1 kilometer
    (c)   1.61 kilometer/1 mile
    (d)   1 kilometer/1.61 mile
    (e)   none of the above

16. How many significant digits are in the unit factor: 1 mile/1.61 kilometer?
    (a)   1
    (b)   2
    (c)   3
    (d)   infinite
    (e)   impossible to determine

17. Which of the following unit factors is derived from: 1 mile = 1760 yards?
    (a)   1 mile/1 yard
    (b)   1760 miles/1 yard
    (c)   1760 yards /1 mile
    (d)   1760 yards/1760 miles
    (e)   none of the above

18. How many significant digits are in the unit factor: 1 mile/1760 yards?
    (a)   1
    (b)   2
    (c)   3
    (d)   infinite
    (e)   impossible to determine

Section 2.9 *Unit Analysis Problem Solving*

19. After carefully reading a problem, what is the *first step* in the unit analysis method
    of problem solving?
    (a)   write down the units asked for in the answer
    (b)   write down the given value related to the answer
    (c)   write one or more unit equations
    (d)   write the two unit factors associated with each unit equation
    (e)   apply a unit factor to convert the given value to the units of the answer

20. After carefully reading a problem, what is the *second step* in the unit analysis
    method of problem solving?
    (a)   write down the units asked for in the answer
    (b)   write down the given value related to the answer
    (c)   write one or more unit equations
    (d)   write the two unit factors associated with each unit equation
    (e)   apply a unit factor to convert the given value to the units of the answer

21. After carefully reading a problem, what is the *third step* in the unit analysis method of problem solving?
    (a) write down the units asked for in the answer
    (b) write down the given value related to the answer
    (c) apply a unit factor to convert the given value to the units of the answer
    (d) enter numbers into your calculator
    (e) none of the above

Section 2.10 *The Percent Concept*

22. A 1¢ coin minted before 1982 is a metal alloy of copper and zinc. If the penny contains 2.869 g copper metal and 0.151 g zinc metal, what is the percent of zinc in the coin?
    (a) 5.00%
    (b) 5.27%
    (c) 15.1%
    (d) 20.0%
    (e) 95.0%

23. A 10¢ coin minted after 1971 is a metal alloy of copper and nickel. If the 10¢ coin has a mass of 2.55 g and is 75.0% copper, what is the mass of nickel in the coin?
    (a) 0.102 g
    (b) 0.340 g
    (c) 0.638 g
    (d) 0.850 g
    (e) 1.91 g

Unit Analysis Problem Solving: Show all work for each of the following.

24. If the Sun is 93,000,000 miles from Earth, how many kilometers is the distance? (Given: One mile equals 1.61 kilometers.)

    Unit Equation:

    Two Unit Factors:

    Unit Analysis Solution:

25. If the Sun is 93,000,000 miles from Earth, how long does it take for sunlight to reach the Earth? (Given: The velocity of light is 186,000 miles per second.)

    Unit Equation:

    Two Unit Factors:

    Unit Analysis Solution:

Chapter 2 Key Terms

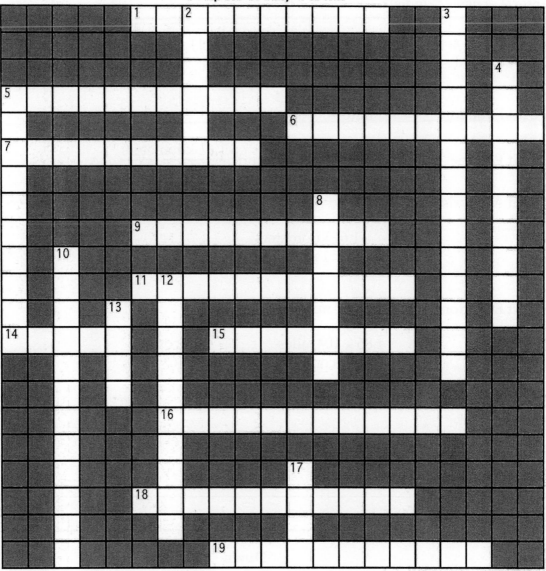

Across

1. 10^n (3 words)
5. eliminating nonsignificant digits (2 words)
6. the type of notation 3.35×10^7
7. a metric unit of length
9. of equal value
11. the digits obtained from a measurement
14. a metric unit of volume
15. a superscript number
16. a systematic problem-solving method (2 words)
18. the inexactness of a measurement
19. a numerical value followed by units

Down

2. the force exerted by gravity on an object
3. the digits often displayed in a calculator
4. a ratio of two equivalent quantities (2 words)
5. the relationship of a fraction and its inverse
8. parts per hundred parts
10. a statement of two equal quantities (2 words)
12. a device for obtaining a measurement
13. a metric unit of mass
17. the quantity of matter in an object

1. e 2. a 3. d 4. b 5. b 6. e 7. d 8. e 9. c 10. d 11. b 12. c 13. b 14. d 15. c 16. c 17. c 18. d 19. a 20. b 21. c 22. a 23. c

24. **Unit Equation:**   1 mile = 1.61 kilometers

Unit Factors: $\dfrac{1 \text{ mile}}{1.61 \text{ kilometers}}$ and $\dfrac{1.61 \text{ kilometers}}{1 \text{ mile}}$

Unit Analysis Solution:

$$9.3 \times 10^7 \ \cancel{\text{miles}} \ \times \ \frac{1.61 \text{ kilometers}}{1 \ \cancel{\text{mile}}} \ = \ 1.5 \times 10^8 \text{ kilometers}$$

25. **Unit Equation:**   186,000 miles = 1 second

Unit Factors: $\dfrac{1.86 \times 10^5 \text{ miles}}{1 \text{ second}}$ and $\dfrac{1 \text{ second}}{1.86 \times 10^5 \text{ miles}}$

Unit Analysis Solution:

$$9.3 \times 10^7 \ \cancel{\text{miles}} \ \times \ \frac{1 \text{ second}}{1.86 \times 10^5 \ \cancel{\text{miles}}} \ = \ 5.0 \times 10^2 \text{ seconds}$$

Chapter 2 Answers to Crossword Puzzle

Across
1. power of ten
5. rounding off
6. scientific
7. centimeter
9. equivalent
11. significant
14. liter
15. exponent
16. unit analysis
18. uncertainty
19. measurement

Down
2. weight
3. nonsignificant
4. unit factor
5. reciprocal
8. percent
10. unit equation
12. instrument
13. gram
17. mass

The Metric System

Section 3.1 *Basic Units and Symbols*

1. Which of the following are the basic units and symbols for the metric system?
    - (a) centimeter (cm), gram (g), liter (L)
    - (b) centimeter (cm), gram (g), milliliter (mL)
    - (c) meter (m), gram (g), liter (L)
    - (d) meter (m), gram (g), liter (l)
    - (e) meter (m), gram (gm), liter (L)

2. What is the symbol for the metric unit microliter?
    - (a) cL
    - (b) mL
    - (c) ML
    - (d) μL
    - (e) none of the above

3. According to the metric system, 1 m = 10 _____.
    - (a) cm
    - (b) dm
    - (c) mm
    - (d) nm
    - (e) μm

4. According to the metric system, 1 g = 100 _____.
    - (a) kg
    - (b) dg
    - (c) cg
    - (d) mg
    - (e) μg

Section 3.2 *Metric Conversion Factors*

5.   Which of the following is a unit factor corresponding to the unit equation:
     1 m = 39.4 in.?
       (a)   1 m/1 in.
       (b)   1 m/39.4 in.
       (c)   39.4 in./39.4 m
       (d)   1 in./39.4 m
       (e)   none of the above

6.   Which of the following is a unit factor corresponding to the unit equation:
     1 m = 100 cm?
       (a)   1 m/1 cm
       (b)   100 m/100 cm
       (c)   1 cm/100 m
       (d)   100 cm/1 m
       (e)   none of the above

Section 3.3 *Metric–Metric Conversions*

7.   What is the three-step sequence, in order, for the unit analysis method of problem
     solving?
       (a)   1–relevant known value, 2–unknown units, 3–unit factor
       (b)   1–unknown units, 2–unit factor, 3–relevant known value
       (c)   1–unknown units, 2–relevant known value, 3–unit factor
       (d)   1–unit factor, 2–unknown units, 3–relevant known value
       (e)   1–unit factor, 2–relevant known value, 3–unknown units

8.   What is the length in meters of a 20.0-mL test tube that measures 15.0 cm?
       (a)   0.150 m
       (b)   0.200 m
       (c)   1.50 m
       (d)   1500 m
       (e)   2000 m

9.   The radius of a tin atom is 1.41×10^{-8} cm. What is the radius of the atom in
     nanometers?
       (a) 1.41×10^{-16} nm
       (b) 1.41×10^{-10} nm
       (c) 1.41×10^{-4} nm
       (d) 1.41×10^{-1} nm
       (e) 1.41×10^{5} nm

Section 3.4 *Metric–English Conversions*

10. Which of the following English–metric equivalents is correct?
    (a) 1 in. = 2.54 cm
    (b) 1 lb = 454 g
    (c) 1 qt = 946 mL
    (d) all of the above
    (e) none of the above

11. A stainless steel cylinder weighs 0.892 lb. What is the mass in kilograms?
    (a) 0.405 kg
    (b) 4.05 kg
    (c) 40.5 kg
    (d) 4050 kg
    (e) 405,000 kg

12. A glass bottle has a volume of 0.750 L. What is the volume in quarts?
    (a) 0.605 qt
    (b) 0.750 qt
    (c) 0.793 qt
    (d) 1.26 qt
    (e) 1.65 qt

Section 3.5 *Volume by Calculation*

13. A piece of gold foil has a volume of 0.645 cm^3. If the piece of gold measures 10.0 cm by 12.5 cm, what is the thickness of the foil?
    (a) 0.000516 cm
    (b) 0.00516 cm
    (c) 0.0516 cm
    (d) 0.516 cm
    (e) 0.806 cm

14. If a subcompact car has an 115-in.^{3} engine, what is the volume of the engine in cubic centimeters?
    (a) 1.88 cm^3
    (b) 7.01 cm^3
    (c) 292 cm^3
    (d) 742 cm^3
    (e) 1880 cm^3

Section 3.6 *Volume by Displacement*

15. What is the term for the method of determining volume by measuring the increase in water level after an object is completely immersed?
    (a) volume by displacement
    (b) volume by difference
    (c) hydrostatic method
    (d) density method
    (e) none of the above

16. A 200.0-g sample of cobalt metal is dropped into a 100-mL graduated cylinder containing 45.5 mL of water. If the resulting water level is 68.0 mL, what is the volume of the cobalt?
    (a) 12.5 mL
    (b) 22.5 mL
    (c) 32.0 mL
    (d) 54.5 mL
    (e) 113.5 mL

Section 3.7 *The Density Concept*

17. If the density of cobalt metal is 4.51 g/mL, which of the following is an associated unit factor?
    (a) 4.51 g/1 mL
    (b) 4.51 mL/1 g
    (c) 4.51 g/4.51 mL
    (d) 1 g/4.51 mL
    (e) 1 g/1 mL

18. If the density of ethyl alcohol is 0.789 g/mL, what is the volume of 35.5 g of ethyl alcohol?
    (a) 2.80 mL
    (b) 4.50 mL
    (c) 28.0 mL
    (d) 45.0 mL
    (e) 280 mL

19. A glass cylinder contains four separate liquid layers: mercury ($d = 13.6$ g/mL), chloroform ($d = 1.49$ g/mL), acetic acid ($d = 1.05$ g/mL), ether ($d = 0.708$ g/mL). If a marble ($d = 3.05$ g/mL) is added to the cylinder, where does it come to rest?
    (a) on top of the mercury layer
    (b) on top of the chloroform layer
    (c) on top of the acetic acid layer
    (d) on top of the ether layer
    (e) on the bottom of the cylinder

Section 3.8 *Temperature*

20. What is the freezing point and boiling point of water on the Celsius scale?
    (a) 0 °C and 100 °C
    (b) 0 °C and 212 °C
    (c) 32 °C and 100 °C
    (d) 32 °C and 212 °C
    (e) 273 °C and 373 °C

21. Given that methanol freezes at −94 °C, what is the freezing point temperature on the Kelvin scale?
    (a) −367 K
    (b) −179 K
    (c) −94 K
    (d) 179 K
    (e) 367 K

Section 3.9 *Heat and Specific Heat*

22. Which of the following may express the total amount of heat energy in a sealed, insulated chamber?
    (a) 25 °C
    (b) 77 °F
    (c) 298 K
    (d) 125 kcal
    (e) all of the above

23. How many kilocalories of heat are required to raise 250.0 g of liquid water from 20.0 °C to 75.0 °C?
    (a) 0.220 kcal
    (b) 4.55 kcal
    (c) 5.00 kcal
    (d) 13.8 kcal
    (e) 18.8 kcal

24. A 50.0-g sample of aluminum releases 420.0 calories when cooled from 100.0 °C to 60.0 °C. What is the specific heat of the metal?
    (a) 0.168 cal/(g × °C)
    (b) 0.210 cal/(g × °C)
    (c) 0.840 cal/(g × °C)
    (d) 4.76 cal/(g × °C)
    (e) 21.0 cal/(g × °C)

Unit Analysis Problem Solving   Show all work for each of the following.

25. If the radius of a nickel atom is 0.125 nm, what is the atomic radius of a nickel atom in picometers (pm)? (Given: $1 \text{ m} = 1 \times 10^{12} \text{ pm}$)

26. If a roll of nickel coins has a mass of 205 g, what is the mass of the roll of nickels in ounces? (Given: 1 lb = 16 ounces)

Chapter 3 Key Terms

Across

1. water freezes at 0° on this temperature scale
5. a method for determining volume
7. the flow of energy from a hot to cold object
9. mass per unit volume
11. a problem-solving method (2 words)
15. a volume equal to one milliliter (2 words)
19. a decimal system of measurement
20. 946 of these volume units equal 1 qt
21. 2.54 of these length units equal 1 in.
22. a statement of two equivalent quantities (2 words)
23. ice melts at 273 on this temperature scale

Down

2. the exact volume of 1 kg of water at 4 °C
3. water freezes at 32° on this temperature scale
4. a mass equal to 1/454 pound
6. a unitless expression related to density (2 words)
8. a metric unit equal to one nutritional Calorie
10. the heat required to raise 1 g of water 1 °C
11. the ratio of two equivalent quantities (2 words)
12. a system of measurement with seven base units
13. a mass approximately equal to two pounds
14. value for water is 1.00 cal/(g × °C) (2 words)
16. a system of measurement with no basic units
17. the average energy of molecules in motion
18. a length approximately equal to a yard

Chapter 3 Answers to Self-Test

1. c 2. d 3. b 4. c 5. b 6. d 7. c 8. a 9. d 10. d 11. a 12. c 13. b 14. e 15. a
16. b 17. a 18. d 19. a 20. a 21. d 22. d 23. d 24. b

25. **Unit Analysis Solution:**

$$0.125 \ \cancel{nm} \ \times \ \frac{1 \ \cancel{m}}{1 \times 10^9 \ \cancel{nm}} \ \times \ \frac{1 \times 10^{12} \ pm}{1 \ \cancel{m}} \ = \ 125 \ pm$$

26. **Unit Analysis Solution:**

$$205 \ \cancel{g} \ \times \ \frac{1 \ \cancel{lb}}{454 \ \cancel{g}} \ \times \ \frac{16 \ oz}{1 \ \cancel{lb}} \ = \ 7.22 \ oz$$

(Note: A summary of *Weights and Measures* including metric and English equivalents is found in Appendix A of the textbook.)

Chapter 3 Answers to Crossword Puzzle

Across
1. Celsius
5. displacement
7. heat
9. density
11. unit analysis
15. cubic centimeter
19. metric
20. mL
21. cm
22. unit equation
23. Kelvin

Down
2. liter
3. Fahrenheit
4. gram
6. specific gravity
8. kilocalorie
10. calorie
11. unit factor
12. SI
13. kilogram
14. specific heat
16. English
17. temperature
18. meter

Matter and Energy

Section 4.1 *Physical States of Matter*

1. Which physical state demonstrates the greatest motion of individual particles?
    (a) solid state
    (b) liquid state
    (c) gaseous state
    (d) the motion of particles cannot be predicted for a physical state
    (e) the motion of particles is the same for each physical state

2. Which of the following describes a substance in the solid physical state?
    (a) The substance has a fixed shape.
    (b) The substance has a fixed volume.
    (c) Particles have an ordered arrangement.
    (d) Particles vibrate in fixed positions.
    (e) all of the above

Section 4.2 *Elements, Compounds, and Mixtures*

3. Which of the following has indefinite, variable properties and can be separated into two or more pure substances by physical methods?
    (a) compound
    (b) element
    (c) mixture
    (d) all of the above
    (e) none of the above

4. Which of the following has definite, constant properties and can be separated into two or more pure substances by chemical methods?
    (a) compound
    (b) element
    (c) mixture
    (d) all of the above
    (e) none of the above

5. Sterling silver is an alloy of copper and silver with the percentage of copper always equal to 7.50%. Classify sterling silver as one of the following.
   (a) compound
   (b) element
   (c) heterogeneous mixture
   (d) homogeneous mixture
   (e) none of the above

Section 4.3 *Names and Symbols of the Elements*

6. Which of the following elements is *not* one of the ten most abundant in the Earth's crust, oceans, and atmosphere?
   (a) hydrogen
   (b) oxygen
   (c) silicon
   (d) titanium
   (e) uranium

Section 4.4 *Metals, Nonmetals, and Semimetals*

7. Which of the following is a general characteristic of a nonmetal?
   (a) brittle in the solid state
   (b) nonconductor of heat
   (c) low density
   (d) low melting point
   (e) all of the above

8. Which of the following is a property of a metal?
   (a) dull appearance
   (b) reacts with nonmetals
   (c) reacts with other metals
   (d) electrical insulator
   (e) all of the above

9. Which of the following elements is an example of a semimetal?
   (a) H
   (b) C
   (c) Fe
   (d) Ge
   (e) Sn

10. Use a periodic table and predict which of the following elements is a gas at 25 °C and 1 atmosphere pressure.
    (a) N
    (b) Na
    (c) Nb
    (d) Nd
    (e) none of the above

11. Use a periodic table and predict which of the following elements is a liquid at 25 °C and 1 atmosphere pressure.
    (a) B
    (b) Ba
    (c) Be
    (d) Br
    (e) none of the above

12. Use a periodic table and predict which of the following elements is a solid at 25 °C and 1 atmosphere pressure.
    (a) H
    (b) He
    (c) Hg
    (d) Ho
    (e) none of the above

Section 4.5 *Compounds and Chemical Formulas*

13. In which of the following compounds are the elements present in a definite proportion by mass?
    (a) H_2O
    (b) CO_2
    (c) N_2O_4
    (d) NaCl
    (e) all compounds have constant composition

14. Ammonium phosphate is used in fertilizer to replenish nitrogen to the soil. If the formula is $(NH_4)_3PO_4$, what is the total number of atoms in one molecule?
    (a) 13
    (b) 16
    (c) 18
    (d) 20
    (e) none of the above

Section 4.6 *Physical and Chemical Properties*

15. Which of the following is an example of a physical property?
    (a) color
    (b) crystalline form
    (c) melting point
    (d) physical state
    (e) all of the above

16. Which of the following is an example of a chemical property?
    (a) density
    (b) ductility
    (c) electrical conductivity
    (d) flammability
    (e) hardness

17. Which of the following is an example of a chemical property of methyl alcohol?
    (a) methyl alcohol and sulfuric acid yield methyl ether
    (b) methyl alcohol in animals produces blindness
    (c) methyl alcohol and sodium metal release hydrogen gas
    (d) methyl alcohol and formic acid give rum flavor
    (e) all of the above

Section 4.7 *Physical and Chemical Changes*

18. Which of the following observations is *not* evidence for a physical change?
    (a) condensation
    (b) crystallization
    (c) deposition
    (d) explosion
    (e) sublimation

19. Which of the following observations is *not* evidence for a chemical change?
    (a) producing bubbles after mixing solutions
    (b) giving a precipitate after mixing solutions
    (c) liberating heat after mixing solutions
    (d) changing color after mixing solutions
    (e) dissolving a solute in solution

Section 4.8 *Conservation of Mass*

20. If 558 kg of powdered iron react with powdered sulfur to produce 879 kg of iron sulfide, what is the mass of reacting sulfur?
    (a) 321 kg
    (b) 558 kg
    (c) 879 kg
    (d) 1437 kg
    (e) impossible to predict from the given information

Section 4.9 *Potential and Kinetic Energy*

21. Which of the following is an example of kinetic energy?
    (a) colliding molecules
    (b) accelerating roller coaster
    (c) swinging in a swing
    (d) rolling wheels
    (e) all of the above

22. Which of the following physical states possesses the greatest attraction between individual particles?
    (a) solid state
    (b) liquid state
    (c) gaseous state
    (d) plasma state
    (e) the attraction is the same for each physical state

23. According to the kinetic theory, temperature measures which of the following?
    (a) molecular attraction
    (b) molecular mass
    (c) molecular motion
    (d) molecular size
    (e) none of the above

Section 4.10 *Conservation of Energy*

24. A nuclear power plant uses radioactive uranium to convert water to steam; steam then drives a turbine that turns a generator for electricity. What are the two forms of energy represented by uranium and the turbine?
    (a) chemical and mechanical energy
    (b) chemical and heat energy
    (c) electrical and mechanical energy
    (d) nuclear and heat energy
    (e) nuclear and mechanical energy

25. According to the law of conservation of mass and energy, which of the following is true for a nuclear fission bomb?
    (a) the mass of the bomb and the fission products are identical
    (b) the energy of the bomb and the fission products are identical
    (c) a small amount of matter is converted into energy
    (d) all of the above
    (e) none of the above

Chemical Symbols of the Elements

26. Provide the chemical symbol for the name of the following elements.

(a) aluminum –

(b) antimony –

(c) argon –

(d) arsenic –

(e) barium –

(f) beryllium –

(g) bismuth –

(h) boron –

(i) bromine –

(j) cadmium –

(k) calcium –

(l) carbon –

(m) chlorine –

(n) chromium –

(o) cobalt –

(p) copper –

(q) fluorine –

(r) germanium –

(s) gold –

(t) helium –

(u) hydrogen –

(v) iodine –

(w) iron –

(x) krypton –

27. Write the name of the element for each of the following chemical symbols.

(a) Pb –

(b) Li –

(c) Mg –

(d) Mn –

(e) Hg –

(f) Ne –

(g) Ni –

(h) N –

(i) O –

(j) P –

(k) Pt –

(l) K –

(m) Ra –

(n) Se –

(o) Si –

(p) Ag –

(q) Na –

(r) Sr –

(s) S –

(t) Te –

(u) Sn –

(v) Ti –

(w) Xe –

(x) Zn –

Chapter 4 Key Terms

Across

1. a change that does not alter composition
8. an element that reacts with metals
9. matter having a constant composition
10. a shiny, high-density element
11. a property that involves another substance
15. a mixture with variable properties
18. it cannot be broken down by chemical reaction
20. the energy of a moving particle
21. a change of state from a gas to a liquid
22. the two related quantities in $E = mc^2$ (3 words)

Down

2. abbreviation for the name of an element
3. all substances have this type of composition
4. a metalloid
5. a change of state from a liquid to a gas
6. a chemical reaction is a chemical _____
7. it cannot be created nor destroyed
9. a direct change of state from a solid to a gas
10. the property of being hammered into a foil
11. it can be broken down into elements
12. the stored energy of chemical composition
13. abbreviation for the name of a compound
14. the property of being drawn into a wire
15. a mixture with consistent properties
16. a physical or chemical characteristic
17. it can be solid, liquid, or gas
19. it can be chemical, mechanical, or electrical

1. c 2. e 3. c 4. a 5. d 6. e 7. e 8. b 9. d 10. a 11. d 12. d 13. e 14. d 15. e
16. d 17. e 18. d 19. e 20. a 21. e 22. a 23. c 24. e 25. c

26. Symbols of Chemical Elements
    Refer to the inside front cover of the textbook and Table 4.3.

27. Names of Chemical Elements
    Refer to the inside front cover of the textbook and Table 4.3.

Chapter 4 Answers to Crossword Puzzle

Across
   1. physical
   8. nonmetal
   9. substance
   10. metal
   11. chemical
   15. heterogeneous
   18. element
   20. kinetic
   21. condensation
   22. mass and energy

Down
   2. symbol
   3. constant
   4. semimetal
   5. vaporization
   6. change
   7. mass
   9. sublimation
   10. malleable
   11. compound
   12. potential
   13. formula
   14. ductile
   15. homogeneous
   16. property
   17. state
   19. energy

Models of the Atom

Section 5.1 *Dalton Model of the Atom*

1. Which of Dalton's atomic theory proposals was later proven to be incorrect?
   (a) An element is composed of tiny particles called atoms.
   (b) All atoms of the same element have the same mass.
   (c) Atoms of different elements combine to form compounds.
   (d) Compounds contain atoms in small whole-number ratios.
   (e) Atoms can combine in more than one whole-number ratio.

Section 5.2 *Thomson Model of the Atom*

2. What do the raisins represent in the plum-pudding analogy of the atom?
   (a) electrons
   (b) neutrons
   (c) protons
   (d) nuclei
   (e) none of the above

3. Where is the mass of an atom found according to the plum-pudding model?
   (a) the mass is found in the electrons
   (b) the mass is found in the protons
   (c) the mass is found in the nucleus
   (d) the mass is distributed homogeneously
   (e) none of the above

Section 5.3 *Rutherford Model of the Atom*

4. What are the approximate diameters of an atom and its nucleus?
    (a) 10^8 cm and 10^{-13} cm, respectively
    (b) 10^{-8} cm and 10^{-13} cm, respectively
    (c) 10^{-13} cm and 10^8 cm, respectively
    (d) 10^{-13} cm and 10^{-8} cm, respectively
    (e) 1 cm and 0.1 cm, respectively

5. State the subatomic particle having a relative charge of zero and an approximate mass of one atomic mass unit.
    (a) alpha
    (b) electron
    (c) neutron
    (d) positron
    (e) proton

Section 5.4 *Atomic Notation*

6. Using atomic notation, indicate the isotope having 10 p$^+$, 12 n^o, and 10 e$^-$.
    (a) $^{12}_{10}\text{Ne}$
    (b) $^{22}_{10}\text{Ne}$
    (c) $^{22}_{12}\text{Ne}$
    (d) $^{12}_{10}\text{Mg}$
    (e) $^{22}_{12}\text{Mg}$

7. How many neutrons are in the nucleus of one atom of oxygen-18?
    (a) 2
    (b) 8
    (c) 10
    (d) 18
    (e) 26

Section 5.5 *Atomic Mass*

8. Nitrogen occurs naturally as ^{14}N and ^{15}N. Which isotope is most abundant? (Hint: Refer to the Periodic Table.)
    (a) nitrogen-7
    (b) nitrogen-8
    (c) nitrogen-14
    (d) nitrogen-15
    (e) none of the above

9. Element **X** occurs naturally as **X**-6 (6.015 amu) and **X**-7 (7.016 amu). Calculate the atomic mass of **X** given that the natural abundance of **X**-7 is 92.5%.
    (a) 6.09 amu
    (b) 6.50 amu
    (c) 6.52 amu
    (d) 6.94 amu
    (e) 12.5 amu

10. Refer to the periodic table and determine the atomic mass of barium.
    (a) 25 amu
    (b) 56 amu
    (c) 81 amu
    (d) 137.33 amu
    (e) 193.33 amu

11. Refer to the periodic table and determine the mass number for an important isotope of radon.
    (a) 86
    (b) 136
    (c) 222
    (d) 308
    (e) none of the above

Section 5.6 *The Wave Nature of Light*

12. Which of the following colors of light has the lowest energy?
    (a) blue
    (b) red
    (c) violet
    (d) yellow
    (e) all colors have the same energy

13. Which of the following colors of light has the shortest wavelength?
    (a) blue
    (b) red
    (c) violet
    (d) yellow
    (e) all colors have the same wavelength

14. Which of the following wavelengths of light has the lowest energy?
    (a) 440 nm
    (b) 470 nm
    (c) 540 nm
    (d) 650 nm
    (e) all wavelengths have the same energy

15. Which of the following wavelengths of light has the lowest frequency?
    (a) 440 nm
    (b) 470 nm
    (c) 540 nm
    (d) 650 nm
    (e) all wavelengths have the same frequency

16. Which of the following wavelengths of light has the fastest velocity?
    (a) 440 nm
    (b) 470 nm
    (c) 540 nm
    (d) 650 nm
    (e) all wavelengths have the same velocity

Section 5.7 *The Quantum Concept*

17. Which of the following instruments gives a quantized measurement?
    (a) 100-mm metric ruler
    (b) 100-g platform balance
    (c) 100-cc hypodermic syringe
    (d) 100-mL volumetric pipet
    (e) none of the above

18. Which of the following is evidence for a quantized change?
    (a) a line spectrum
    (b) a continuous spectrum
    (c) a visible spectrum
    (d) a rainbow
    (e) all of the above

Section 5.8 *Bohr Model of the Atom*

19. Which of these statements is *not* true of the Bohr model of the atom?
    (a) Electrons maintain a constant energy state as they orbit the nucleus.
    (b) Electrons that receive heat energy jump to a higher orbit.
    (c) Electrons gain energy if they drop to an orbit closer to the nucleus.
    (d) Electrons circle the nucleus similar to the way planets circle the sun.
    (e) none of the above

20. Which of the following produces the "atomic fingerprint" for hydrogen?
    (a) atoms of helium reacting
    (b) atoms of helium colliding
    (c) electrons jumping to a higher energy level
    (d) electrons dropping to a lower energy level
    (e) electrons being captured by the nucleus

21. How many photons of light are emitted when the electron in a hydrogen atom drops from energy level 5 to 2?
    (a) 1
    (b) 2
    (c) 3
    (d) 5
    (e) none of the above

Section 5.9 *Energy Levels and Sublevels*

22. How many energy sublevels theoretically exist in the 5th energy level?
    (a) 2
    (b) 5
    (c) 10
    (d) 18
    (e) none of the above

23. What is the maximum number of electrons that can theoretically occupy the 5*d* sublevel?
    (a) 2
    (b) 5
    (c) 6
    (d) 10
    (e) none of the above

24. What is the maximum number of electrons that can occupy the 2nd energy level?
    (a) 2
    (b) 6
    (c) 8
    (d) 10
    (e) 18

Section 5.10 *Electron Configuration*

25. What element has the following electron configuration: $1s^2\ 2s^2\ 2p^6\ 3s^2\ 3p^1$?
    (a) B
    (b) Na
    (c) Al
    (d) K
    (e) Sc

26. What element has the following electron configuration: $1s^2\ 2s^2\ 2p^6\ 3s^2\ 3p^3$?
    (a) N
    (b) Al
    (c) P
    (d) K
    (e) Sc

Section 5.11 *Quantum Mechanical Model of the Atom*

27. Which of the following is a distinction between an orbit and an orbital?
    (a) An orbital represents an energy level.
    (b) An orbital represents a probability boundary.
    (c) An orbital may contain more than one electron.
    (d) An orbital may not gain or lose energy.
    (e) An orbit and an orbital are identical.

28. Which of the following orbitals has a dumbbell shape in the 5th energy level?
    (a) 5s
    (b) 4p
    (c) 5p
    (d) 4d
    (e) 5d

29. What is the maximum number of electrons that can occupy a 4d orbital?
    (a) 2
    (b) 3
    (c) 4
    (d) 6
    (e) 10

30. How many orbitals exist within the 4d sublevel?
    (a) 2
    (b) 4
    (c) 5
    (d) 6
    (e) 10

31. What is the maximum number of electrons that can occupy a 4d sublevel?
    (a) 2
    (b) 4
    (c) 7
    (d) 10
    (e) 14

32. What is the maximum number of electrons that can occupy the 4th energy level?
    (a) 2
    (b) 4
    (c) 10
    (d) 18
    (e) 32

Chapter 5 Key Terms

Across

1. the distance light travels in one cycle
3. a spectrum showing a broad band of radiant energy
7. an electron orbit designated 1, 2, 3 ... (2 words)
8. the arrangement of electrons in an atom
10. a spectrum with narrow bands of radiant energy
14. a region of high density in the center of the atom
16. an atomic model with electron probability (2 words)
20. the _____ number refers to the A value
23. a formula for a wavelength of light from a H atom
24. a model of an atom with electron orbits
25. a neutral subatomic particle
26. the number of wave cycles in one second

Down

2. the light spectrum from 400–700 nm
4. instrument for viewing the lines in a spectrum
5. a region of high probability for finding an electron
6. expresses the composition of a nucleus (2 words)
9. 1/12 the mass of a carbon-12 atom
11. atoms that differ only by the number of neutrons
12. the _____ number refers to the Z value
13. a spectrum from X rays to microwaves (2 words)
15. a discrete energy level in an atom
17. the principle that the precise location and energy of an electron cannot both be determined
18. a negatively charged subatomic particle
19. the weighted average mass of isotopes (2 words)
21. an energy level that is designated s, p, d, f ...
22. a positively charged subatomic particle

1. **b** 2. **a** 3. **d** 4. **b** 5. **c** 6. **b** 7. **c** 8. **c** 9. **d** 10. **d** 11. **c** 12. **b** 13. **c** 14. **d** 15. **d** 16. **e**
17. **d** 18. **a** 19. **c** 20. **d** 21. **a** 22. **b** 23. **d** 24. **c** 25. **c** 26. **c** 27. **b** 28. **c** 29. **a** 30. **c**
31. **d** 32. **e**

Chapter 5 Answers to Crossword Puzzle

Across
1. wavelength
3. continuous
7. energy level
8. configuration
10. line
14. nucleus
16. quantum mechanical
20. mass
23. Balmer
24. Bohr
25. neutron
26. frequency

Down
2. visible
4. spectroscope
5. orbital
6. atomic notation
9. amu
11. isotopes
12. atomic
13. radiant energy
15. quantum
17. uncertainty
18. electron
19. atomic mass
21. sublevel
22. proton

The Periodic Table

Section 6.1 *Classification of Elements*

1. Which of the following scientists did *not* contribute to the development of the modern periodic table of elements?
    (a) Johann Döbereiner
    (b) Albert Einstein
    (c) Dmitri Mendeleev
    (d) H.G.J. Moseley
    (e) J.A.R. Newlands

2. Why did Mendeleev *not* include germanium (Ge) in his periodic table of 1871?
    (a) Ge is unstable.
    (b) Ge is radioactive.
    (c) Ge had not been discovered in 1871.
    (d) Ge is found only in compounds.
    (e) none of the above

Section 6.2 *The Periodic Law Concept*

3. The original periodic law was based upon which of the following?
    (a) increasing atomic mass
    (b) increasing atomic number
    (c) increasing isotopic mass
    (d) increasing mass number
    (e) increasing neutron number

4. The modern periodic law is based on which of the following?
    (a) increasing atomic mass
    (b) increasing atomic number
    (c) increasing isotopic mass
    (d) increasing mass number
    (e) increasing neutron number

Section 6.3 *Groups and Periods of Elements*

5. Which fourth period representative element has the lowest atomic number?
    (a) C
    (b) K
    (c) Sc
    (d) Ga
    (e) Y

6. Which fourth period transition element has the lowest atomic number?
    (a) C
    (b) K
    (c) Sc
    (d) Ga
    (e) Y

7. Previous to the 1985 recommendation by IUPAC, what was the designation for the nitrogen family of elements according to the American convention?
    (a) IIIA
    (b) IIIB
    (c) VA
    (d) VB
    (e) none of the above

Section 6.4 *Periodic Trends*

8. Which of the following is a general trend from left to right in the periodic table?
    (a) atomic radius increases; metallic character increases
    (b) atomic radius increases; metallic character decreases
    (c) atomic radius decreases; metallic character increases
    (d) atomic radius decreases; metallic character decreases
    (e) none of the above

9. Predict which of the following elements has the largest atomic radius.
    (a) F
    (b) Cl
    (c) O
    (d) S
    (e) Se

10. Predict which of the following elements has the most nonmetallic character.
    (a) F
    (b) Cl
    (c) O
    (d) S
    (e) Se

Section 6.5 *Properties of Elements*

11. Predict the density of barium, given the density of calcium (1.54 g/cm^3) and strontium (2.63 g/cm^3).
    (a) 0.45 g/cm^3
    (b) 1.09 g/cm^3
    (c) 2.09 g/cm^3
    (d) 3.72 g/cm^3
    (e) 4.17 g/cm^3

12. Predict the atomic radius of antimony given the radius of arsenic (0.125 nm) and bismuth (0.155 nm).
    (a) 0.030 nm
    (b) 0.095 nm
    (c) 0.140 nm
    (d) 0.185 nm
    (e) 0.280 nm

13. Predict the boiling point for argon given the boiling points of krypton ($-152 \,°\text{C}$) and xenon ($-107 \,°\text{C}$).
    (a) $-197 \,°\text{C}$
    (b) $-130 \,°\text{C}$
    (c) $-62 \,°\text{C}$
    (d) $-45 \,°\text{C}$
    (e) $45 \,°\text{C}$

14. Predict which of the following elements has chemical properties most similar to iron.
    (a) Al
    (b) Ru
    (c) Na
    (d) Ni
    (e) Zn

Section 6.6 *Blocks of Elements*

15. Which energy sublevel is being filled by the elements Ga through Kr?
    (a) $3d$
    (b) $4s$
    (c) $4p$
    (d) $4d$
    (e) $4f$

16. Which energy sublevel is being filled by the elements Y through Cd?
    (a) $4d$
    (b) $5s$
    (c) $5p$
    (d) $5d$
    (e) $5f$

17. Which of the following is the electron configuration for an atom of tin?
    (a) [Kr] $5s^2 4d^{10} 4p^2$
    (b) [Kr] $5s^2 4d^{10} 5p^2$
    (c) [Kr] $5s^2 4d^{10} 5d^2$
    (d) [Xe] $5s^2 4d^{10} 5p^2$
    (e) [Xe] $5s^2 4d^{10} 5d^2$

Section 6.7 *Valence Electrons*

18. Predict the number of valence electrons for a Group IIA/2 element.
    (a) 1
    (b) 2
    (c) 3
    (d) 6
    (e) 8

19. Predict the number of valence electrons for an atom of selenium.
    (a) 2
    (b) 4
    (c) 6
    (d) 16
    (e) 34

Section 6.8 *Electron Dot Formulas*

20. Which of the following is the electron dot formula for an atom of strontium?
    (a) Sr·   (b) Sr·   (c)·Sr: (d) ·Sr: (e) :Sr:

21. Which of the following is the electron dot formula for an atom of oxygen?
    (a) O·   (b) O·   (c) ·O·   (d) ·O: (e) :O:

Section 6.9 *Ionization Energy*

22. Which of the following groups of elements has the highest ionization energy?
    (a) Group IA/1
    (b) Group IIA/2
    (c) Group IIB/12
    (d) Group VIIA/17
    (e) Group VIIIA/18

23. Which of the following elements has the lowest ionization energy?
    (a) Rb
    (b) Sr
    (c) Cs
    (d) Ba
    (e) Xe

24. Which of the following is a general trend for the ionization energy of elements in the periodic table?
    (a) increases from left to right; increases from bottom to top
    (b) increases from left to right; decreases from bottom to top
    (c) decreases from left to right; increases from bottom to top
    (d) decreases from left to right; decreases from bottom to top
    (e) none of the above

Section 6.10 *Ionic Charges*

25. Which of the following groups has a predictable ionic charge of two negative?
    (a) Group IIA/2
    (b) Group IVB/4
    (c) Group IVA/14
    (d) Group VIA/16
    (e) Group VIIIA/17

26. Which of the following ions is *not* isoelectronic with argon?
    (a) S^{2-}
    (b) Cl^-
    (c) K^+
    (d) Sc^{3+}
    (e) V^{4+}

27. What is the core electron configuration for a titanium(II) ion, Ti^{2+}?
    (a) [Ar]
    (b) [Ar] $4s^2$
    (c) [Ar] $3d^2$
    (d) [Ar] $4s^2\ 3d^2$
    (e) none of the above

28. What is the core electron configuration for a gallium ion, Ga^{3+}?
    (a) [Ar]
    (b) [Ar] $3d^{10}$
    (c) [Ar] $4s^2\ 3d^8$
    (d) [Ar] $4s^2\ 3d^{10}\ 4p^1$
    (e) none of the above

29. What is the core electron configuration for a fluoride ion, F⁻?
    (a) [Ne]
    (b) [Ne] $2p^5$
    (c) [Ne] $2p^6$
    (d) [Ne] $3s^1$
    (e) none of the above

30. What is the core electron configuration for a telluride ion, Te²⁻?
    (a) [Kr]
    (b) [Kr] $4p^6$
    (c) [Xe]
    (d) [Xe] $5p^6$
    (e) none of the above

Periodic Table Exercise

Refer to a periodic table and identify the element by its chemical symbol(s) that matches each of the following descriptions.

    (a) the lowest atomic number alkali metal
    (b) the lowest atomic number alkaline earth metal
    (c) the lowest atomic number halogen
    (d) the lowest atomic number noble gas
    (e) the lowest atomic number lanthanide
    (f) the lowest atomic number actinide
    (g) the lowest atomic number rare earth element
    (h) the lowest atomic number transuranium element
    (i) the radioactive alkali metal
    (j) the radioactive alkaline earth metal
    (k) the radioactive lanthanide
    (l) the radioactive rare earth element
    (m) the Group IA/1 nonmetal
    (n) the Group IIIA/13 semimetal
    (o) the Group IVA/14 nonmetal
    (p) the Group IB/11 coinage metals

Chapter 6 Key Terms

Across

2. an element such as F, Cl, Br, I
5. an element such as Sc, Y, La (2 words)
6. properties of elements repeat in a pattern (2 words)
7. charge on an atom after gaining or losing electrons
9. the elements with variable chemical properties
11. Lewis structure (2 words)
14. symbol of a noble gas element
17. the elements filling 4*f* and 5*f* sublevels (2 words)
22. symbol of an element more valuable than gold
23. symbol of the least dense metallic element
24. an atom that has gained or lost electrons
25. symbol of a radioactive Group 1 element
26. a row of elements
27. symbol of an alkaline earth element
28. the nucleus and inner electrons of an atom

Down

1. symbol of a liquid nonmetallic element
3. an element with an atomic number 90–103
4. the unreactive gaseous elements
5. the elements with predictable chemical properties
8. ions with identical electron configurations
10. the metals Mg, Ca, Sr (2 words)
12. the elements following atomic number 92
13. symbol of the most dense metallic element
15. an element with an atomic number 58–71
16. a family of elements
18. the metals Li, Na, K
19. the process of an atom losing an electron
20. the outer *s* and *p* electrons
21. a simplified notation for electron configuration

Chapter 6 Answers to Self-Test

1. **b** 2. **c** 3. **a** 4. **b** 5. **b** 6. **c** 7. **c** 8. **d** 9. **e** 10. **a** 11. **d** 12. **c** 13. **a** 14. **b** 15. **c** 16. **a**
17. **b** 18. **b** 19. **c** 20. **b** 21. **d** 22. **e** 23. **c** 24. **a** 25. **d** 26. **e** 27. **c** 28. **b** 29. **a** 30. **c**

Periodic Table Exercise

(a) Li
(b) Be
(c) F
(d) He
(e) Ce
(f) Th
(g) Sc
(h) Np
(i) Fr
(j) Ra
(k) Pm
(l) Pm
(m) H
(n) B
(o) C
(p) Cu, Ag, Au

Chapter 6 Answers to Crossword Puzzle

Across
2. halogen
5. rare earth
6. periodic law
7. ionic
9. transition
11. electron dot
14. Ar
17. inner transition
22. Pt
23. Li
24. ion
25. Fr
26. period
27. Mg
28. kernel

Down
1. Br
3. actinide
4. noble
5. representative
8. isoelectronic
10. alkaline earth
12. transuranium
13. Os
15. lanthanide
16. group
18. alkali
19. ionization
20. valence
21. core

Language of Chemistry

Section 7.1 *Classification of Compounds*

1. The compound $CaSO_4$ is classified as which of the following?
   - (a) binary ionic
   - (b) ternary ionic
   - (c) binary molecular
   - (d) binary acid
   - (e) ternary oxyacid

2. The compound H_2O is classified as which of the following?
   - (a) binary ionic
   - (b) ternary ionic
   - (c) binary molecular
   - (d) binary acid
   - (e) ternary oxyacid

3. Aqueous HBr is classified as which of the following?
   - (a) binary ionic
   - (b) ternary ionic
   - (c) binary molecular
   - (d) binary acid
   - (e) ternary oxyacid

4. The ammonium ion, NH_4^+, is classified as which of the following?
   - (a) monoatomic cation
   - (b) monoatomic anion
   - (c) polyatomic cation
   - (d) polyatomic anion
   - (e) none of the above

5. The hydroxide ion, OH^-, is classified as which of the following?
   - (a) monoatomic cation
   - (b) monoatomic anion
   - (c) polyatomic cation
   - (d) polyatomic anion
   - (e) none of the above

Section 7.2 Monoatomic Ions

6. What is the systematic IUPAC name for Hg_2^{2+} according to the Stock system?
    - (a) mercury ion
    - (b) mercury(I) ion
    - (c) mercury(II) ion
    - (d) mercuric ion
    - (e) mercurous ion

7. What is the systematic IUPAC name for Cu^{2+} according to the Latin system?
    - (a) copper ion
    - (b) copper(II) ion
    - (c) cupric ion
    - (d) cuprous ion
    - (e) cuprum ion

8. What is the systematic IUPAC name for I^-?
    - (a) hypoiodite ion
    - (b) iodide ion
    - (c) iodite ion
    - (d) iodate ion
    - (e) periodate ion

9. What is the predicted ionic charge of an element in Group VIA/16?
    - (a) 2+
    - (b) 6+
    - (c) 2–
    - (d) 6–
    - (e) none of the above

Section 7.3 Polyatomic Ions

10. What is the systematic IUPAC name for MnO_4^-?
    - (a) manganese ion
    - (b) manganate ion
    - (c) manganite ion
    - (d) permanganate ion
    - (e) none of the above

11. What is the formula for the hydrogen carbonate ion?
    - (a) $C_2H_3O_2^-$
    - (b) $C_2H_3O_2^{2-}$
    - (c) HCO_3^-
    - (d) HCO_3^{2-}
    - (e) none of the above

Section 7.4 *Writing Chemical Formulas*

12. What is the chemical formula for the ionic compound composed of calcium ions, Ca^{2+}, and phosphide ions, P^{3-}?
    (a) Ca_2P_2
    (b) Ca_2P_3
    (c) Ca_3P_2
    (d) Ca_3P_3
    (e) Ca_6P_6

13. What is the chemical formula for the ionic compound composed of strontium ions, Sr^{2+} and phosphate ions, PO_4^{3-}?
    (a) $SrPO_4$
    (b) $Sr_2(PO_4)_3$
    (c) $Sr_3(PO_4)_2$
    (d) $Sr_6(PO_4)_6$
    (e) none of the above

14. What is the chemical formula for the ionic compound composed of barium ions, Ba^{2+}, and cyanide ions, CN^-?
    (a) $BaCN$
    (b) $BaCN_2$
    (c) Ba_2CN
    (d) $BaCN)_2$
    (e) $Ba_2(CN)_2$

Section 7.5 *Binary Ionic Compounds*

15. What is the ionic charge for the cobalt ion in Co_2S_3?
    (a) 2+
    (b) 2–
    (c) 3+
    (d) 3–
    (e) none of the above

16. What is the systematic IUPAC name for Co_2S_3 according to the Stock system?
    (a) cobalt sulfide
    (b) cobalt(II) sulfide
    (c) cobalt(II) sulfite
    (d) cobalt(III) sulfide
    (e) cobalt(III) sulfite

17. Predict the chemical formula for rubidium fluoride, given the formula of sodium fluoride, NaF.
    (a) RbF
    (b) Rb_2F
    (c) RbF_2
    (d) Rb_2F_3
    (e) Rb_3F_2

Section 7.6 *Ternary Ionic Compounds*

18. What is the ionic charge for the cobalt ion in $Co_3(PO_4)_2$?
    (a) 2+
    (b) 2–
    (c) 3+
    (d) 3–
    (e) none of the above

19. What is the systematic IUPAC name for $Co_3(PO_4)_2$ according to the Latin system?
    (a) cobaltic phosphide
    (b) cobaltous phosphite
    (c) cobaltic phosphite
    (d) cobaltous phosphate
    (e) cobaltic phosphate

20. Predict the chemical formula for aluminum selenide, given the formula of aluminum oxide, Al_2O_3.
    (a) AlSe
    (b) Al_2Se
    (c) $AlSe_3$
    (d) Al_2Se_3
    (e) Al_3Se_2

Section 7.7 *Binary Molecular Compounds*

21. What is the systematic IUPAC name for SF_6?
    (a) sulfur tetrafluoride
    (b) sulfur hexafluoride
    (c) sulfur heptafluoride
    (d) sulfur octafluoride
    (e) sulfur nonafluoride

22. What is the formula for tetraphosphorus trisulfide?
    (a) PS_3
    (b) P_4S
    (c) P_3S_4
    (d) P_4S_3
    (e) P_4S_4

Section 7.8 *Binary Acids*

23. What is the systematic IUPAC name for aqueous HBr?
    (a) hydrogen bromide
    (b) hydrobromous acid
    (c) hydrobromic acid
    (d) bromous acid
    (e) bromic acid

24. What is the formula for hydrosulfuric acid?
    (a) $H_2S(aq)$
    (b) $HSO_3(aq)$
    (c) $HSO_4(aq)$
    (d) $H_2SO_3(aq)$
    (e) $H_2SO_4(aq)$

Section 7.9 *Ternary Oxyacids*

25. What is the systematic IUPAC name for aqueous $HC_2H_3O_2$?
    (a) acetic acid
    (b) carbonic acid
    (c) dicarbonic acid
    (d) hydrogen carbonous acid
    (e) hydrogen carbonic acid

26. What is the formula for sulfuric acid?
    (a) $H_2S(aq)$
    (b) $HSO_3(aq)$
    (c) $HSO_4(aq)$
    (d) $H_2SO_3(aq)$
    (e) $H_2SO_4(aq)$

Chemical Nomenclature Exercise

27. Give an acceptable IUPAC name for each of the following chemical formulas.

   (a) KF

   (b) CuO

   (c) $Zn(NO_3)_2$

   (d) $NiSO_4$

   (e) $(NH_4)_2CrO_4$

28. Provide the chemical formula for each of the following compounds.

   (a) silver iodide

   (b) cadmium nitride

   (c) manganese(II) chloride

   (d) tin(IV) carbonate

   (e) iron(II) hydroxide

Chapter 7 Key Terms

Across

3. a polyatomic anion containing oxygen
4. any positively charged ion
6. classification of aqueous H_2SO_4 (2 words)
7. the simplest particle in an ionic compound (2 words)
8. classification of aqueous HCl (2 words)
10. a compound of two nonmetals (2 words)
13. a multiple-atom ion
15. -ous and -ic suffix naming system
17. releases hydrogen ions in water
18. Roman numeral naming system

Down

1. releases hydroxide ions in water
2. a substance not containing carbon
5. any negatively charged ion
6. a compound of a metal and two nonmetals (2 words)
8. a compound of a metal and one nonmetal (2 words)
9. an ionic compound from an acid–base reaction
11. a solution of a substance dissolved in water
12. a single-atom ion
14. the simplest particle in a molecular compound
16. International Union of Pure and Applied Chemistry

Chapter 7 Answers to Self-Test

1. **b** 2. **c** 3. **d** 4. **c** 5. **d** 6. **b** 7. **c** 8. **b** 9. **c** 10. **d** 11. **c** 12. **c** 13. **c** 14. **d** 15. **c** 16. **d** 17. **a** 18. **a** 19. **d** 20. **d** 21. **b** 22. **d** 23. **c** 24. **a** 25. **a** 26. **e**

Chemical Nomenclature Exercise

27. (a) potassium fluoride
    (b) copper(II) oxide, or cupric oxide
    (c) zinc nitrate
    (d) nickel(II) sulfate
    (e) ammonium chromate

28. (a) AgI
    (b) Cd_3N_2
    (c) $MnCl_2$
    (d) $Sn(CO_3)_2$
    (e) $Fe(OH)_2$

(Note: To assist you in learning the names and the formulas of ions, flashcards are available at the back of this *Study Guide & Selected Solutions Manual.*)

Chapter 7 Answers to Crossword Puzzle

Across

3. oxyanion
4. cation
6. ternary oxyacid
7. formula unit
8. binary acid
10. binary molecular
13. polyatomic
15. Latin
17. acid
18. Stock

Down

1. base
2. inorganic
5. anion
6. ternary ionic
8. binary ionic
9. salt
11. aqueous
12. monoatomic
14. molecule
16. IUPAC

Chemical Reactions

Section 8.1 *Evidence for Chemical Reactions*

1. Which of the following is evidence for a chemical reaction?
    (a) a gas is detected
    (b) a precipitate is formed
    (c) a color change is observed
    (d) an energy change is noted
    (e) all of the above

2. Which of the following is *not* evidence for a chemical reaction producing a gas?
    (a) an odor is detected
    (b) a color change is observed
    (c) a flaming splint is extinguished
    (d) a glowing splint bursts into flames
    (e) bubbles are observed

Section 8.2 *Writing Chemical Equations*

3. Which of the statements below best describes the following reaction?

$$2\,AgClO_3(s) \xrightarrow{\Delta} 2\,AgCl(s) + 3\,O_2(g)$$

    (a) Silver chlorate gives silver chloride and oxygen.
    (b) Silver chlorate is heated to give silver chloride and oxygen.
    (c) Silver chlorate gives solid silver chloride and oxygen gas.
    (d) Silver chlorate is heated to give solid silver chloride and oxygen gas.
    (e) Solid silver chlorate is heated to give solid silver chloride and oxygen gas.

Section 8.3 *Balancing Chemical Equations*

4. Which of the following is *not* a general direction for balancing an equation?
    (a) write correct formulas for reactants and products
    (b) begin balancing with the most complex formula
    (c) balance polyatomic ions as a single unit
    (d) balance ionic compounds as a single unit
    (e) check each reactant and product to verify the coefficients

5. What is the coefficient of carbon dioxide after balancing the following equation?

$$\_CuHCO_3(s) \xrightarrow{\Delta} \_Cu_2CO_3(s) + \_H_2O(g) + \_CO_2(g)$$

    (a) 1   (b) 2   (c) 3   (d) 4   (e) 6

6. What is the coefficient of water after balancing the following equation?

$$\_HNO_3(aq) + \_Co(OH)_2(aq) \rightarrow \_Co(NO_3)_2(aq) + \_H_2O(l)$$

    (a) 1   (b) 2   (c) 3   (d) 4   (e) 6

7. What is the coefficient of gold(III) sulfate after balancing the following equation?

$$\_ZnSO_4(aq) + \_Au(NO_3)_3(aq) \rightarrow \_Au_2(SO_4)_3(s) + \_Zn(NO_3)_2(aq)$$

    (a) 1   (b) 2   (c) 3   (d) 4   (e) 6

8. What is the coefficient of lead metal after balancing the following equation?

$$\_Al(s) + \_Pb(C_2H_3O_2)_2(aq) \rightarrow \_Al(C_2H_3O_2)_3(aq) + \_Pb(s)$$

    (a) 1   (b) 2   (c) 3   (d) 4   (e) 6

Section 8.4 *Classifying Chemical Reactions*

9. What type of chemical reaction is illustrated in Exercise 8?
    (a) combination reaction
    (b) decomposition reaction
    (c) single-replacement reaction
    (d) double-replacement reaction
    (e) neutralization reaction

Section 8.5 *Combination Reactions*

10. What is the product predicted from the following combination reaction?

$$Li(s) \quad + \quad O_2(g) \quad \overset{\Delta}{\rightarrow}$$

   (a) LiO
   (b) Li_2O
   (c) LiO_2
   (d) Li_2O_3
   (e) Li_3O_2

11. What is the product predicted from the following combination reaction?

$$S(s) \quad + \quad O_2(g) \quad \overset{\Delta}{\rightarrow}$$

   (a) SO
   (b) S_2O
   (c) SO_2
   (d) SO_3
   (e) It is impossible to predict a single product for a nonmetal and oxygen without more information.

12. What is the product predicted from the following combination reaction?

$$Ni(s) \quad + \quad Br_2(g) \quad \overset{\Delta}{\rightarrow}$$

   (a) NiBr
   (b) Ni_2Br
   (c) $NiBr_2$
   (d) Ni_2Br_3
   (e) Ni_3Br_2

Section 8.6 *Decomposition Reactions*

13. What are the products from the following decomposition reaction?

$$Bi(HCO_3)_3(s) \quad \overset{\Delta}{\rightarrow}$$

   (a) Bi, H_2O, and CO_2
   (b) Bi_2CO_3, H_2, and CO_2
   (c) $Bi_2(CO_3)_3$ and H_2O
   (d) Bi_2CO_3, H_2O, and CO_2
   (e) $Bi_2(CO_3)_3$, H_2O, and CO_2

14. What are the products from the following decomposition reaction?

$$Co_2(CO_3)_3(s) \xrightarrow{\Delta}$$

    (a)   Co and CO_2
    (b)   CoO and CO_2
    (c) Co_2O_3 and CO_2
    (d)   CoO, H_2O, and CO_2
    (e) Co_2O_3, H_2O, and CO_2

15. What are the products from the following decomposition reaction?

$$Ca(ClO_3)_2(s) \xrightarrow{\Delta}$$

    (a)   Ca and CO_2
    (b)   Ca, Cl_2, and O_2
    (c) $CaCl_2$ and H_2O
    (d) $CaCl_2$ and O_2
    (e) $CaCl_2$ and CO_2

Section 8.7 *The Activity Series Concept*

16. Which of the following metals reacts with aqueous $FeSO_4$?
Partial Activity Series: Mg > Al > Zn > Fe > (H)
    (a)   Al
    (b)   Mg
    (c)   Zn
    (d)   all of the above
    (e)   none of the above

17. Which of the following metals reacts with aqueous $AgNO_3$?
Partial Activity Series: (H) > Cu > Ag > Hg > Au
    (a)   Au
    (b)   Cu
    (c)   Hg
    (d)   all of the above
    (e)   none of the above

18. Which of the following metals does *not* react with sulfuric acid?
Partial Activity Series: Mn > Cd > Co > (H) > Ag
    (a)   Ag
    (b)   Cd
    (c)   Co
    (d)   Mn
    (e)   Sn

19. Which of the following metals reacts with water at 25 °C?
    (a) Al
    (b) Fe
    (c) Zn
    (d) Mg
    (e) Sr

Section 8.8 *Single-Replacement Reactions*

20. What are the products from the following single-replacement reaction?

$$Mg(s) \ + \ AgNO_3(aq) \quad \rightarrow$$

    (a) Ag and $Mg(NO_3)_2$
    (b) Ag and $Mg(NO_2)_2$
    (c) Ag_2O and $Mg(NO_3)_2$
    (d) Ag_2O and $Mg(NO_2)_2$
    (e) no reaction

21. What are the products from the following single-replacement reaction?

$$Zn(s) \ + \ H_2SO_4(aq) \quad \rightarrow$$

    (a) ZnO and H_2SO_3
    (b) ZnO and H_2S
    (c) $ZnSO_4$ and H_2
    (d) $ZnSO_4$ and H_2O
    (e) no reaction

22. What are the products from the following single-replacement reaction?

$$Na(s) \ + \ H_2O(l) \quad \rightarrow$$

    (a) Na_2O and H_2
    (b) Na_2O and H_2O
    (c) NaOH and H_2
    (d) NaOH and H_2O
    (e) no reaction

Section 8.9 *Solubility Rules*

23. Which of the following solid compounds is insoluble in water?
    (Refer to the Solubility Rules in Appendix E of the textbook.)
    (a) $(NH_4)_2CO_3$
    (b) K_2CrO_4
    (c) $PbSO_4$
    (d) Na_2S
    (e) $Sr(OH)_2$

Section 8.10 *Double-Replacement Reactions*

24. What are the products from the following double-replacement reaction?

$$BaCl_2(aq) \quad + \quad K_2SO_4(aq) \quad \rightarrow$$

    (a) BaS and $KClO_4$
    (b) $BaSO_3$ and KCl
    (c) $BaSO_3$ and $KClO_4$
    (d) $BaSO_4$ and KCl
    (e) $BaSO_4$ and $KClO_4$

Section 8.11 *Neutralization Reactions*

25. What are the products from the following neutralization reaction?

$$HNO_3(aq) \quad + \quad NH_4OH(aq) \quad \rightarrow$$

    (a) NH_3NO_3 and H_2O
    (b) NH_3NO_3 and O_2
    (c) NH_4NO_3 and H_2O
    (d) NH_4NO_3 and O_2
    (e) NH_4NO_2 and H_2O

Writing Chemical Equations for Chemical Reactions

26. Write a balanced chemical equation for the reaction of aqueous strontium nitrate with aqueous sodium sulfate to yield a precipitate of strontium sulfate and aqueous sodium nitrate.

27. Write a balanced chemical equation for the reaction of hydrochloric acid with aqueous barium hydroxide to give aqueous barium chloride and water.

Chapter 8 Key Terms

Across

2. an insoluble salt in solution
5. a type of reaction involving a single reactant
7. a number in a chemical formula
9. a compound produced from an acid–base reaction
10. releases hydrogen ions in solution
11. a relative order of metals to react (2 words)
15. a substance that speeds up a chemical reaction
16. a type of replacement reaction that switches ions
18. a solution with a substance dissolved in water
19. a reaction requires this substance

Down

1. a chemical change
3. a reaction that absorbs heat
4. a reaction yields this substance
6. a type of reaction involving an acid and base
8. a type of reaction involving two elements
9. a type of replacement reaction of a metal in an acid
10. an element that reacts with water (2 words)
12. a number preceding a formula in an equation
13. a description of a reaction using formulas and symbols
14. a reaction that releases heat
17. releases hydroxide ions in solution

Chapter 8 Answers to Self-Test

1. **e** 2. **b** 3. **e** 4. **d** 5. **a** 6. **b** 7. **a** 8. **c** 9. **c** 10. **b** 11. **e** 12. **c** 13. **e** 14. **c** 15. **d** 16. **d** 17. **b** 18. **a** 19. **e** 20. **a** 21. **c** 22. **c** 23. **c** 24. **d** 25. **c**

Translating Chemical Reactions into Balanced Equations

26. $Sr(NO_3)_2(aq) + Na_2SO_4(aq) \rightarrow SrSO_4(s) + 2\,NaNO_3(aq)$

27. $2\,HCl(aq) + Ba(OH)_2(aq) \rightarrow BaCl_2(aq) + 2\,HOH(l)$

Chapter 8 Answers to Crossword Puzzle

Across
2. precipitate
5. decomposition
7. subscript
9. salt
10. acid
11. activity series
15. catalyst
16. double
18. aqueous
19. reactant

Down
1. reaction
3. endothermic
4. product
6. neutralization
8. combination
9. single
10. active metal
12. coefficient
13. equation
14. exothermic
17. base

The Mole Concept

Section 9.1 *Avogadro's Number*

1. How many atoms of carbon have a mass equal to its atomic mass, 12.01 g?
    (a)   1
    (b)   6
    (c)   12
    (d)   12.01
    (e) 6.02×10^{23}

2. What is the mass of Avogadro's number of iron atoms?
    (a)   55.85 amu
    (b)   55.85 grams
    (c) 6.02×10^{23} amu
    (d) 6.02×10^{23} grams
    (e) 55.85×10^{23} grams

Section 9.2 *Mole Calculations I*

3. Which of the following is equal to 1 mol of substance?
    (a) 6.02×10^{23} sodium atoms, Na
    (b) 6.02×10^{23} chlorine molecules, Cl_2
    (c) 6.02×10^{23} sodium chloride formula units, NaCl
    (d) 6.02×10^{23} chloride ions, Cl^-
    (e)   all of the above

4. How many xenon atoms are in 0.250 mol of the noble gas?
    (a) 1.51×10^{22} atoms
    (b) 1.51×10^{23} atoms
    (c) 1.51×10^{24} atoms
    (d) 2.41×10^{23} atoms
    (e) 2.41×10^{24} atoms

5. How many moles of argon correspond to 7.52×10^{22} atoms of the inert gas?
    (a) 0.0125 mol
    (b) 0.0801 mol
    (c) 0.125 mol
    (d) 0.801 mol
    (e) 1.25 mol

Section 9.3 *Molar Mass*

6. What is the approximate molar mass of trinitrotoluene (TNT), $C_7H_5(NO_2)_3$?
    (a) 63.07 g/mol
    (b) 135.13 g/mol
    (c) 189.21 g/mol
    (d) 227.15 g/mol
    (e) 405.39 g/mol

Section 9.4 *Mole Calculations II*

7. What is the mass of 4.50×10^{22} atoms of gold?
    (a) 0.0679 g
    (b) 0.0748 g
    (c) 13.3 g
    (d) 14.7 g
    (e) 197 g

8. How many methane molecules, CH_4, have a mass of 3.20 g?
    (a) 1.20×10^{23} molecules
    (b) 1.20×10^{24} molecules
    (c) 1.93×10^{24} molecules
    (d) 3.01×10^{23} molecules
    (e) 3.01×10^{24} molecules

9. What is the mass of a formula unit of cuprous sulfide, Cu_2S?
    (a) 1.04×10^{-26} g
    (b) 1.66×10^{-24} g
    (c) 2.64×10^{-22} g
    (d) 3.79×10^{21} g
    (e) 9.57×10^{25} g

Section 9.5 *Molar Volume*

10. One mole of which of the following gases occupies 22.4 L at STP?
    - (a) ammonia, NH_3
    - (b) helium, He
    - (c) hydrogen, H_2
    - (d) oxygen, O_2
    - (e) all of the above

11. What is the density of nitrogen gas, N_2, at STP?
    - (a) 0.625 g/L
    - (b) 0.800 g/L
    - (c) 1.25 g/L
    - (d) 1.60 g/L
    - (e) 22.4 g/L

12. If the density of nitrous oxide is 1.96 g/L at STP, what is the molar mass of laughing gas?
    - (a) 11.4 g/mol
    - (b) 19.6 g/mol
    - (c) 22.4 g/mol
    - (d) 43.9 g/mol
    - (e) 196 g/mol

Section 9.6 *Mole Calculations III*

13. What is the volume of 1.51×10^{24} molecules of ammonia gas, NH_3, at STP?
    - (a) 0.112 L
    - (b) 0.399 L
    - (c) 2.51 L
    - (d) 8.93 L
    - (e) 56.2 L

14. What is the mass of 0.500 liter of oxygen gas, O_2, at STP?
    - (a) 0.286 g
    - (b) 0.714 g
    - (c) 3.50 g
    - (d) 6.40 g
    - (e) 112 g

Section 9.7 *Percent Composition*

15. The formula for the illegal drug cocaine is $C_{17}H_{21}NO_4$. What is the percentage of nitrogen in the compound?
    (a) 4.62%
    (b) 6.99%
    (c) 21.1%
    (d) 32.6%
    (e) 67.3%

16. Phenacyl chloride is used as tear gas for riot control and has the formula C_7H_7OCl. What is the percentage of carbon in the compound?
    (a) 4.96%
    (b) 8.42%
    (c) 11.2%
    (d) 24.9%
    (e) 59.0%

Section 9.8 *Empirical Formula*

17. If 0.250 mol V reacts with 0.375 mol O, what is the empirical formula of vanadium oxide?
    (a) VO
    (b) V_2O_3
    (c) V_2O_5
    (d) V_3O_2
    (e) V_5O_2

18. If 1.500 g of vanadium metal react with oxygen gas to give 2.679 g of vanadium oxide, what is the empirical formula of the product?
    (a) VO
    (b) V_2O_3
    (c) V_2O_5
    (d) V_3O_2
    (e) V_5O_2

19. Acetaldehyde is used in the manufacture of perfumes, flavors, and dyes. Calculate the empirical formula for acetaldehyde given that its percent composition is: 54.53% C, 9.15% H, and 36.32% O.
    (a) $C_1H_1O_1$
    (b) $C_2H_4O_1$
    (c) $C_2H_3O_1$
    (d) $C_5H_9O_2$
    (e) $C_6H_9O_3$

Section 9.9 *Molecular Formula*

20. Find the molecular formula for aspirin given that its empirical formula is $C_9H_8O_4$ and its approximate molar mass is 175 g/mol.
    (a) $C_1H_1O_1$
    (b) $C_2H_2O_1$
    (c) $C_9H_8O_4$
    (d) $C_{18}H_{16}O_8$
    (e) $C_{27}H_{24}O_{12}$

21. Lindane is an insecticide that causes dizziness and diarrhea in humans. Find the molecular formula for lindane given that its percent composition is: 24.8% C, 2.1% H, and 73.1% Cl. The approximate molar mass of lindane is 290 g/mol.
    (a) $C_1H_1Cl_1$
    (b) $C_3H_3Cl_3$
    (c) $C_1H_1Cl_6$
    (d) $C_6H_6Cl_6$
    (e) $C_{12}H_1Cl_{35}$

Analogies for Avogadro's Number

22. Given 6.02×10^{23} hydrogen atoms (0.07 nm in diameter), placed side-by-side, which of the following is the best estimate of the total length?
    (a) a diameter of a pin
    (b) a centimeter
    (c) a meter stick
    (d) a kilometer
    (e) a million trips around Earth's equator

23. Given 6.02×10^{23} steel 16-pound shotput balls, which of the following is the best estimate of their total mass?
    (a) a microgram
    (b) a kilogram
    (c) an automobile
    (d) the Earth
    (e) the universe

24. Given 6.02×10^{23} softballs, which of the following is the best estimate of their total volume?
    (a) 1000-mL beaker
    (b) the Los Angeles Coliseum
    (c) the Grand Canyon
    (d) the Earth
    (e) the universe

Empirical and Molecular Formulas Show all work for each of the following.

25. Galactose and glucose comprise milk sugar, lactose. If the percent composition of galactose is 40.00% C, 6.72% H, 53.29% O, what is its empirical formula?

26. If the approximate molar mass of galactose is about 180 g/mol, what is the molecular formula of galactose?

Chapter 9 Key Terms

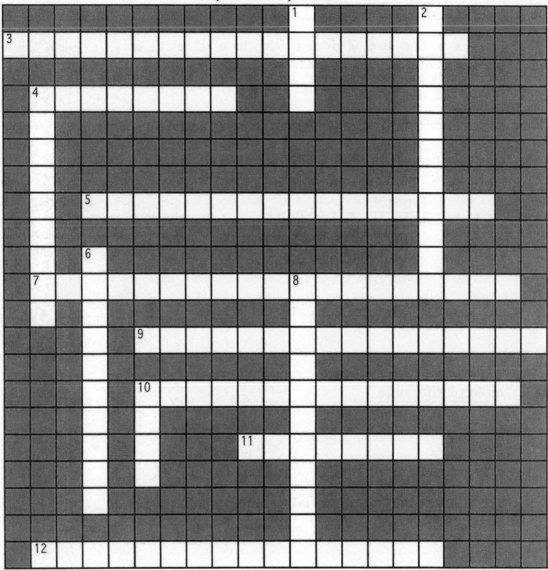

Across

3. the amount of each element in a compound (2 words)
4. particle in a molecular compound
5. the simplest ratio of atoms in a molecule (2 words)
7. 0 °C (2 words)
9. the actual ratio of atoms in a molecule (2 words)
10. number of atoms in 12.01 g of carbon (2 words)
11. stated equal volumes of gas have equal molecules
12. 1 atmosphere (2 words)

Down

1. amount of substance with 6.02×10^{23} particles
2. particle in an ionic compound (2 words)
4. the mass of 1 mol of substance (2 words)
6. usually expressed in grams per liter (2 words)
8. the space occupied by 1 mol of gas (2 words)
10. representative particle of an element

1. e   2. b   3. e   4. b   5. c   6. d   7. d   8. a   9. c   10. e   11. c   12. d   13. e   14. b   15. a
16. e   17. b   18. c   19. b   20. c   21. d   22. e   23. d   24. d

Empirical and Molecular Formulas

25. $40.00 \, \text{g C} \times \dfrac{1 \, \text{mol C}}{12.01 \, \text{g C}} = 3.330 \, \text{mol C}$

   $6.72 \, \text{g H} \times \dfrac{1 \, \text{mol H}}{1.01 \, \text{g H}} = 6.65 \, \text{mol H}$

   $53.29 \, \text{g O} \times \dfrac{1 \, \text{mol O}}{16.00 \, \text{g O}} = 3.330 \, \text{mol O}$

   thus, empirical formula is $C_{\frac{3.33}{3.33}} H_{\frac{6.65}{3.33}} O_{\frac{3.33}{3.33}} = C_{1.00}H_{1.99}O_{1.00} = CH_2O$

26. $\dfrac{\text{molecular formula}}{\text{empirical formula}} = \dfrac{(CH_2O)_n}{CH_2O}$

   $n = \dfrac{180 \, \text{g}}{30 \, \text{g}} = 6$

   thus, molecular formula is $(CH_2O)_6 = C_6H_{12}O_6$

Chapter 9 Answers to Crossword Puzzle

Across
3. percent composition
4. molecule
5. empirical formula
7. standard temperature
9. molecular formula
10. Avogadro's number
11. Avogadro
12. standard pressure

Down
1. mole
2. formula unit
4. molar mass
6. gas density
8. molar volume
10. atom

Chemical Equation Calculations

Section 10.1 *Interpreting a Chemical Equation*

1. How many moles of oxygen gas, O_2, react with 2 moles of nitrogen monoxide gas, NO, according to the following equation?

$$\_NO(g) \ + \ \_O_2(g) \ \xrightarrow{\Delta/UV} \ \_NO_2(g)$$

   (a) 1 mol
   (b) 2 mol
   (c) 3 mol
   (d) 4 mol
   (e) none of the above

2. How many molar masses of nitrogen dioxide gas are produced from 2 molar masses of nitrogen monoxide gas according to the following equation?

$$\_NO(g) \ + \ \_O_2(g) \ \xrightarrow{\Delta/UV} \ \_NO_2(g)$$

   (a) 1 MM
   (b) 2 MM
   (c) 3 MM
   (d) 4 MM
   (e) none of the above

3. How many liters of oxygen gas react to produce 2 L of nitrogen dioxide gas (at similar conditions) according to the following equation?

$$\_NO(g) \ + \ \_O_2(g) \ \xrightarrow{\Delta/UV} \ \_NO_2(g)$$

   (a) 1 L
   (b) 2 L
   (c) 3 L
   (d) 4 L
   (e) none of the above

4. In an experiment 0.300 g of NO gas reacts to produce 0.460 g of NO_2 gas. Using the conservation of mass law, predict the mass of reacting oxygen gas.

$$\_NO(g) \ + \ \_O_2(g) \quad \xrightarrow{\Delta/UV} \quad \_NO_2(g)$$

(a) 0.0800 g
(b) 0.160 g
(c) 0.320 g
(d) 0.380 g
(e) 0.760 g

Section 10.2 Mole–Mole Problems

5. How many moles of water react with 0.500 mol of lithium metal according to the following reaction?

$$\_Li(s) \ + \ \_H_2O(l) \quad \rightarrow \quad \_LiOH(aq) \ + \ \_H_2(g)$$

(a) 0.125 mol
(b) 0.250 mol
(c) 0.500 mol
(d) 1.00 mol
(e) 2.00 mol

6. How many moles of hydrogen gas are produced from the reaction of 0.500 mol of sodium metal?

$$\_Na(s) \ + \ \_H_2O(l) \quad \rightarrow \quad \_NaOH(aq) \ + \ \_H_2(g)$$

(a) 0.125 mol
(b) 0.250 mol
(c) 0.500 mol
(d) 1.00 mol
(e) 2.00 mol

7. How many moles of water must react in order to produce 0.500 mol of potassium hydroxide?

$$\_K(s) \ + \ \_H_2O(l) \quad \rightarrow \quad \_KOH(aq) \ + \ \_H_2(g)$$

(a) 0.125 mol
(b) 0.250 mol
(c) 0.500 mol
(d) 1.00 mol
(e) 2.00 mol

Section 10.3 *Types of Stoichiometry Problems*

8. Classify the following type of stoichiometry problem: *How many grams of sulfur must react with an excess volume of oxygen in order to yield 0.500 L of sulfur dioxide gas?*
    - (a) mass–mass problem
    - (b) mass–volume problem
    - (c) volume–volume problem
    - (d) mole–mole
    - (e) none of the above

9. Classify the following type of stoichiometry problem: *How many grams of sulfur must react with an excess volume of oxygen in order to yield 0.500 g of sulfur dioxide gas?*
    - (a) mass–mass problem
    - (b) mass–volume problem
    - (c) volume–volume problem
    - (d) mole–mole
    - (e) none of the above

10. Classify the following type of stoichiometry problem: *How many liters of sulfur must react with an excess volume of oxygen in order to yield 50.0 cm³ of sulfur dioxide gas?*
    - (a) mass–mass problem
    - (b) mass–volume problem
    - (c) volume–volume problem
    - (d) mole–mole
    - (e) none of the above

11. In general, how many unit factors are necessary to solve mass–mass stoichiometry problems?
    - (a) one
    - (b) two
    - (c) three
    - (d) four
    - (e) five

12. In general, how many unit factors are necessary to solve mass–volume stoichiometry problems?
    - (a) one
    - (b) two
    - (c) three
    - (d) four
    - (e) five

13. Which of the following steps is *not* necessary to solve a volume–volume stoichiometry problem?
    - (a) write a balanced chemical equation for the reaction
    - (b) calculate the moles of known gas given its volume
    - (c) convert the volume of known gas to the volume of unknown gas
    - (d) determine the volume of known gas
    - (e) determine the volume of unknown gas

Section 10.4 *Mass–Mass Problems*

14. How many grams of iron are produced from the reaction of 500.0 g of aluminum metal?

$$\_\_FeO(l) \ + \ \_\_Al(l) \ \xrightarrow{\Delta} \ \_\_Fe(l) \ + \ \_\_Al_2O_3(s)$$

(a) 345 g
(b) 689 g
(c) 1030 g
(d) 1550 g
(e) 3100 g

15. How many grams of aluminum metal must react to give 500.0 g of iron?

$$\_\_FeO(l) \ + \ \_\_Al(l) \ \xrightarrow{\Delta} \ \_\_Fe(l) \ + \ \_\_Al_2O_3(s)$$

(a) 80.6 g
(b) 161 g
(c) 242 g
(d) 362 g
(e) 483 g

Section 10.5 *Mass–Volume Problems*

16. What STP volume of oxygen is released from the decomposition of 2.166 g of mercuric oxide (216.59 g/mol)?

$$\_\_HgO(s) \ \xrightarrow{\Delta} \ \_\_Hg(l) \ + \ \_\_O_2(g)$$

(a) 0.112 L
(b) 0.224 L
(c) 0.448 L
(d) 1.12 L
(e) 2.24 L

17. How many grams of mercuric oxide (216.59 g/mol) must decompose to release 0.375 L of oxygen gas at STP?

$$\_\_HgO(s) \ \xrightarrow{\Delta} \ \_\_Hg(l) \ + \ \_\_O_2(g)$$

(a) 0.0752 g
(b) 1.82 g
(c) 3.63 g
(d) 7.25 g
(e) 14.5 g

Section 10.6 *Volume–Volume Problems*

18. Calculate the volume of oxygen gas that reacts with 32.0 mL of sulfur dioxide. (Assume constant conditions.)

$$\underline{\quad}SO_2(g) \ + \ \underline{\quad}O_2(g) \ \xrightarrow[]{650\ °C} \ \underline{\quad}SO_3(g)$$

    (a)   8.00 mL
    (b)   16.0 mL
    (c)   32.0 mL
    (d)   64.0 mL
    (e)   96.0 mL

19. Calculate the volume of nitrogen gas that produces 50.0 mL of ammonia, NH_3. (Assume constant conditions.)

$$\underline{\quad}N_2(g) \ + \ \underline{\quad}H_2(g) \ \xrightarrow[]{500\ °C/500\ atm} \ \underline{\quad}NH_3(g)$$

    (a)   6.25 mL
    (b)   12.5 mL
    (c)   25.0 mL
    (d)   50.0 mL
    (e)   100.0 mL

Section 10.7 *The Limiting Reactant Concept*

20. If 1.00 mol of chromium reacts with 1.00 mol of oxygen gas according to the following equation, how many moles of chromium(III) oxide are produced?

$$\underline{\quad}Cr(s) \ + \ \underline{\quad}O_2(g) \ \xrightarrow[]{\Delta} \ \underline{\quad}Cr_2O_3(s)$$

    (a)   0.500 mol
    (b)   0.667 mol
    (c)   1.00 mol
    (d)   1.50 mol
    (e)   2.00 mol

Section 10.8 *Limiting Reactant Problems*

21. If 20.0 mL of methane gas reacts with 20.0 mL of oxygen gas, what is the volume of carbon dioxide gas produced (assume constant conditions)?

$$\underline{\quad}CH_4(g) \ + \ \underline{\quad}O_2(g) \ \xrightarrow[]{spark} \ \underline{\quad}CO_2(g) \ + \ \underline{\quad}H_2O(g)$$

    (a)   5.00 mL
    (b)   10.0 mL
    (c)   20.0 mL
    (d)   40.0 mL
    (e)   80.0 mL

22. If 10.0 g of aluminum metal react with 10.0 g of sulfur according to the following equation, how many grams of aluminum sulfide are produced?

$$\_\_Al(s) \quad + \quad \_\_S(s) \quad \overset{\Delta}{\longrightarrow} \quad \_\_Al_2S_3(s)$$

    (a)   15.6 g
    (b)   20.0 g
    (c)   27.8 g
    (d)   46.8 g
    (e)   55.7 g

Section 10.9 *Percent Yield*

23. Starting with 0.657 g of lead(II) nitrate, a student collects 0.905 g of precipitate. If the calculated mass of precipitate is 0.914 g, what is the percent yield?
    (a)   71.9%
    (b)   72.6%
    (c)   99.0%
    (d)   101%
    (e)   138%

Stoichiometry Problems Show all work for each of the following.

24. What is the mass of solid silver chloride produced from the reaction of excess aqueous silver nitrate and 0.333 g of barium chloride?

25. What is the volume of oxygen gas liberated at STP when 1.555 g of solid silver chlorate is decomposed with heat to give solid silver chloride?

Chapter 10 Key Terms

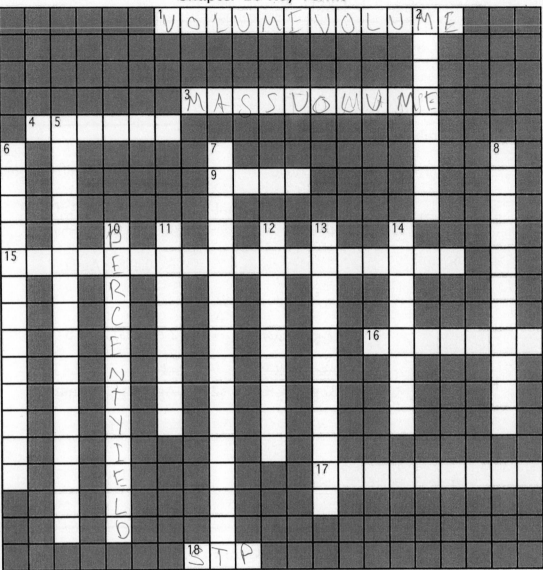

Across

1. liters-to-liters type of problem (2 words)
3. grams-to-liters type of problem (2 words)
4. the experimental yield from a reaction
9. the element produced from the blast furnace process
15. Lavoisier's law (3 words)
16. the substance produced from the Haber process
17. ratio of moles of reactants and products (2 words)
18. 0 °C and 1 atm

Down

2. a single particle composed of nonmetal atoms
5. Gay-Lussac's law (2 words)
6. relates quantities according to a balanced equation
7. controls the maximum amount of product (2 words)
8. the calculated yield from a reaction
10. ratio of actual to theoretical yield (2 words)
11. stated equal volumes—same number of molecules
12. the mass of one mole of substance (2 words)
13. the volume of one mole of gas (2 words)
14. grams-to-grams type of problem (2 words)

1. **a** 2. **b** 3. **a** 4. **b** 5. **c** 6. **b** 7. **c** 8. **b** 9. **a** 10. **c** 11. **c** 12. **c** 13. **b** 14. **d** 15. **b**
16. **a** 17. **d** 18. **b** 19. **c** 20. **a** 21. **b** 22. **a** 23. **c**

Stoichiometry Problems

24. $$BaCl_2(s) \ + \ 2\,AgNO_3(aq) \ \rightarrow \ 2\,AgCl(s) \ + \ Ba(NO_3)_2(aq)$$

$$0.333 \ \cancel{g\ BaCl_2} \times \frac{1 \ \cancel{mol\ BaCl_2}}{208.23 \ \cancel{g\ BaCl_2}} \times \frac{2 \ \cancel{mol\ AgCl}}{1 \ \cancel{mol\ BaCl_2}} \times \frac{143.32 \ g\ AgCl}{1 \ \cancel{mol\ AgCl}} \ = \ g\ AgCl$$

$$= \ 0.458 \ g\ AgCl$$

25. $$2\,AgClO_3(s) \ \overset{\Delta}{\rightarrow} \ 2\,AgCl(s) \ + \ 3\,O_2(g)$$

$$1.555 \ \cancel{g\ AgClO_3} \times \frac{1 \ \cancel{mol\ AgClO_3}}{191.32 \ \cancel{g\ AgClO_3}} \times \frac{3 \ \cancel{mol\ O_2}}{2 \ \cancel{mol\ AgClO_3}} \times \frac{22.4 \ L\ O_2}{1 \ \cancel{mol\ O_2}} \ = \ L\ O_2$$

$$= \ 0.273 \ L\ O_2$$

Across
1. volume–volume
3. mass–volume
4. actual
9. iron
15. conservation of mass
16. ammonia
17. mole ratio
18. STP

Down
2. molecule
5. combining volumes
6. stoichiometry
7. limiting reactant
8. theoretical
10. percent yield
11. Avogadro
12. molar mass
13. molar volume
14. mass–mass

The Gaseous State

Section 11.1 *Properties of Gases*

1. Which of the following is an observed property of gases?
    (a) gases have a variable shape and volume
    (b) gases expand infinitely
    (c) gases have low density
    (d) gases mix completely
    (e) all of the above

Section 11.2 *Atmospheric Pressure*

2. Which of the following does *not* express standard atmospheric pressure?
    (a) 29.9 in. Hg
    (b) 14.7 psi
    (c) 76 cm Hg
    (d) 760 torr
    (e) 101 Pa

3. If the pressure of nitrogen gas is 15 mm Hg, what is the pressure in atm?
    (a) 0.020 atm
    (b) 0.20 atm
    (c) 15 atm
    (d) 1100 atm
    (e) 11,000 atm

4. If the atmospheric pressure is 30.4 in. Hg, what is the pressure in psi?
    (a) 14.9 psi
    (b) 30.4 psi
    (c) 149 psi
    (d) 3990 psi
    (e) 39,900 psi

Section 11.3 *Variables Affecting Gas Pressure*

5. The pressure exerted by a gas is affected by three factors. Which of the following *increases* the pressure?
    (a) increasing the volume
    (b) decreasing the temperature
    (c) increasing the number of gas molecules
    (d) all of the above
    (e) none of the above

Section 11.4 *Boyle's Law*

6. Which of the following graphs represents the plot of pressure versus the volume at constant temperature?

(a)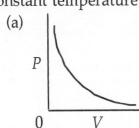

(b)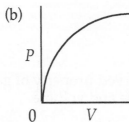

(c)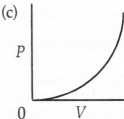

(d)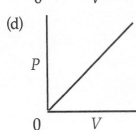

(e)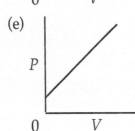

7. A sample of ammonia gas at 1.20 atm compresses from 500 mL to 250 mL. If the temperature remains constant, what is the new pressure of the gas?
    (a) 0.600 atm
    (b) 1.00 atm
    (c) 1.20 atm
    (d) 2.40 atm
    (e) 4.80 atm

Section 11.5 *Charles's Law*

8. A sample of helium gas at 45 °C expands from 125 mL to 250 mL. If the pressure remains constant, what is the new Celsius temperature of the gas?
    (a) 23 °C
    (b) 90 °C
    (c) 159 °C
    (d) 363 °C
    (e) 636 °C

9. Which of the following graphs represents the plot of volume versus the Kelvin temperature at constant pressure?

(a)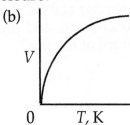

(b)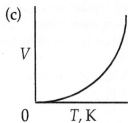

(c)

(d)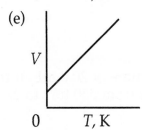

(e)

Section 11.6 *Gay-Lussac's Law*

10. Which of the following graphs represents the plot of pressure versus the Kelvin temperature at constant volume?

(a)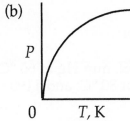

(b)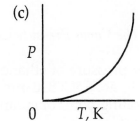

(c)

(d)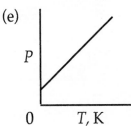

(e)

11. A sample of chlorine gas at 785 mm Hg is heated from 50 °C to 100 °C. If the volume remains constant, what is the new pressure of the gas?
    (a) 153 mm Hg
    (b) 393 mm Hg
    (c) 680 mm Hg
    (d) 907 mm Hg
    (e) 1570 mm Hg

Section 11.7 Combined Gas Law

12. A sample of hydrogen gas is in a cylinder at 0 °C. If the gas is heated so that the pressure increases from 1.00 atm to 2.00 atm, and the volume increases from 2.50 L to 5.00 L, what is the new temperature of the gas?
    (a)   68 K
    (b)   135 K
    (c)   273 K
    (d)   546 K
    (e)   1092 K

13. A sample of oxygen gas has a volume of 20.0 mL. If the gas cools from 450 K to 225 K, and the pressure decreases from 550 torr to 275 torr, what is the new volume of the gas?
    (a)   5.00 mL
    (b)   10.0 mL
    (c)   20.0 mL
    (d)   40.0 mL
    (e)   80.0 mL

Section 11.8 The Vapor Pressure Concept

14. The vapor pressure of ethanol is 360 mm Hg at 60 °C; 600 mm Hg at 72 °C; 760 mm Hg at 78 °C; 810 mm Hg at 80 °C; and 1190 mm Hg at 90 °C. What is the normal boiling point of ethanol?
    (a)   60 °C
    (b)   72 °C
    (c)   78 °C
    (d)   80 °C
    (e)   90 °C

Section 11.9 Dalton's Law

15. An air sample contains nitrogen, oxygen, and argon gases. If the partial pressure of nitrogen is 588 torr, oxygen is 158 torr, and argon is 7 torr, what is the pressure of the air sample?
    (a)   7 torr
    (b)   14 torr
    (c)   423 torr
    (d)   753 torr
    (e)   768 torr

16. Oxygen is collected over water at 50.0 °C and 770.5 mm Hg. What is the partial pressure of the oxygen gas? (Vapor pressure of water at 50 °C is 92.5 mm Hg.)
    (a) 678.0 mm Hg
    (b) 720.5 mm Hg
    (c) 820.5 mm Hg
    (d) 863.0 mm Hg
    (e) none of the above

Section 11.10 *Ideal Gas Behavior*

17. Which of the following is *not* true according to the kinetic theory of gases?
    (a) molecules occupy zero volume
    (b) molecules are attracted to one another
    (c) molecules move randomly
    (d) molecules have elastic collisions
    (e) molecules have a kinetic energy proportional to temperature

18. At the same temperature, which of the following gases has the fastest velocity?
    (a) hydrogen
    (b) helium
    (c) oxygen
    (d) fluorine
    (e) The average velocity for each gas is the same.

19. What is the temperature at which an ideal gas exerts zero pressure?
    (a) 0 °C
    (b) 273 °C
    (c) 0 K
    (d) 273 K
    (e) −273 K

Section 11.11 *Ideal Gas Law*

20. If 1.00 mol of krypton gas exerts a pressure of 1.00 atm at 100 °C, what is the volume of the gas? (R = 0.0821 atm•L/mol•K)
    (a) 2.20 mL
    (b) 8.21 L
    (c) 12.2 L
    (d) 30.6 L
    (e) 1220 L

Combined Gas Law Problems

21. Calcium metal reacts with water to produce hydrogen gas. If 45.5 mL of hydrogen is collected over water at 20 °C and 752 mm Hg, what is the volume at STP? (The vapor pressure of water at 20 °C is 17.5 mm Hg.)

	P	V	T
initial			
final			

22. The decomposition of a baking soda sample by heating gives 5.50 L of carbon dioxide gas at 100 °C and 2.25 atm pressure. If the volume is 2.77 L at 7.50 atm, what is the temperature of the gas?

	P	V	T
initial			
final			

Chapter 11 Key Terms

Across

4. gas law that relates *P*, *V*, and *T*
5. gas that obeys kinetic theory
6. the lowest possible temperature (2 words)
7. *PV* = _____
8. pressure exerted by a gas in a mixture (2 words)
14. relates *P*, *V*, *T*, and moles of gas (3 words)
17. gas that deviates from kinetic theory
18. molecular collisions that do not lose energy
19. 0 °C and 1 atm
20. SI unit of pressure
21. 1 mm Hg of pressure
22. instrument for measuring atmospheric pressure
23. stated the relationship of gas pressure and volume

Down

1. *P* and *V* are _____ proportional
2. gas–liquid equilibrium pressure
3. *T* and *V* are _____ proportional
4. stated the relationship of volume and temperature
6. the pressure exerted by molecules in air
8. abbreviation for an English unit of pressure
9. 76 cm Hg = 1 _____
10. technique for determining the volume of a gas
11. describes the behavior of an ideal gas (2 words)
12. a volume containing no gas molecules
13. stated the relationship of pressure and temperature
15. stated pressure of gases in a mixture is additive
16. frequency of molecular collisions (2 words)

Chapter 11 Answers to Self-Test

1. **e**   2. **e**   3. **a**   4. **a**   5. **c**   6. **a**   7. **d**   8. **d**   9. **d**   10. **d**   11. **d**   12. **e**   13. **c**   14. **c**   15. **d**
16. **a**   17. **b**   18. **a**   19. **c**   20. **d**

Combined Gas Law Problems

21.

	P	V	T
initial	$752 - 17.5 = 734$ mm Hg	45.5 mL	20 °C + 273 = 293 K
final	760 mm Hg	V_{final}	273 K

$$45.5 \text{ mL} \times \frac{734 \text{ mm Hg}}{760 \text{ mm Hg}} \times \frac{273 \text{ K}}{293 \text{ K}} = 40.9 \text{ mL}$$

22.

	P	V	T
initial	2.25 atm	5.50 L	100 °C + 273 = 373 K
final	7.50 atm	2.77 L	T_{final}

$$373 \text{ K} \times \frac{7.50 \text{ atm}}{2.25 \text{ atm}} \times \frac{2.77 \text{ L}}{5.50 \text{ L}} = 626 \text{ K } (353 \text{ °C})$$

Chapter 11 Answers to Crossword Puzzle

Across
 4. combined
 5. ideal
 6. absolute zero
 7. *nRT*
 8. partial pressure
14. ideal gas constant
17. real
18. elastic
19. STP
20. kilopascal
21. torr
22. barometer
23. Boyle

Down
 1. inversely
 2. vapor
 3. directly
 4. Charles
 6. atmospheric
 8. psi
 9. atmosphere
10. displacement
11. kinetic theory
12. vacuum
13. Gay-Lussac
15. Dalton
16. gas pressure

Chemical Bonding

Section 12.1 *The Chemical Bond Concept*

1. What type of chemical bond results from the attraction between a positive metal ion and a negative nonmetal ion?
    (a) covalent bond
    (b) ionic bond
    (c) metallic bond
    (d) valence bond
    (e) none of the above

2. Predict which of the following compounds is held together by covalent bonds.
    (a) CO_2
    (b) PbO_2
    (c) SnO_2
    (d) TiO_2
    (e) all of the above

Section 12.2 *Ionic Bonds*

3. Which of the following statements is true regarding the bond between sodium and chlorine in a NaCl formula unit?
    (a) Sodium atoms lose electrons, and chlorine atoms gain electrons.
    (b) Sodium and chloride ions bond by electrostatic attraction.
    (c) The ionic radius of a sodium ion is less than its atomic radius.
    (d) Breaking an ionic bond between Na^+ and Cl^- requires energy.
    (e) all of the above

4. How many valence electrons are in a bromine atom and a bromide ion?
    (a) 1 and 7, respectively
    (b) 1 and 8, respectively
    (c) 7 and 8, respectively
    (d) 7 and 10, respectively
    (e) none of the above

5. Which noble gas is isoelectronic with a potassium ion?
   (a) neon
   (b) argon
   (c) krypton
   (d) xenon
   (e) none of the above

6. Which of the following ions has the following electron configuration: $1s^2\ 2s^2\ 2p^6\ 3s^2\ 3p^6\ 4s^2\ 3d^{10}\ 4p^6$?
   (a) I^-
   (b) S^{2-}
   (c) K^+
   (d) Sr^{2+}
   (e) none of the above

7. What is the predicted electron configuration for a titanium(IV) ion, Ti^{4+}?
   (a) $1s^2\ 2s^2\ 2p^6\ 3s^2\ 3p^6$
   (b) $1s^2\ 2s^2\ 2p^6\ 3s^2\ 3p^6\ 4s^2$
   (c) $1s^2\ 2s^2\ 2p^6\ 3s^2\ 3p^6\ 4s^2\ 3d^2$
   (d) $1s^2\ 2s^2\ 2p^6\ 3s^2\ 3p^6\ 4s^2\ 3d^6$
   (e) none of the above

Section 12.3 *Covalent Bonds*

8. Which of the following statements is true regarding a covalent bond between hydrogen and fluorine atoms in a HF molecule?
   (a) Valence electrons are transferred from hydrogen to fluorine atoms.
   (b) Bonding electrons are distributed over the entire HF molecule.
   (c) The bond length is equal to the sum of the two atomic radii.
   (d) Breaking a bond between H and F atoms releases energy.
   (e) none of the above

Section 12.4 *Electron Dot Formulas of Molecules*

9. Draw the electron dot formula for silane, SiH_4. How many pairs of nonbonding electrons are in a silane molecule?
   (a) 0
   (b) 1
   (c) 2
   (d) 4
   (e) none of the above

10. Draw the electron dot formula for acetylene, C_2H_4. State the type of bonds in an acetylene molecule.
    (a) 5 single bonds
    (b) 4 single bonds and 1 triple bond
    (c) 4 single bonds and 1 double bond
    (d) 2 double bonds and 1 triple bond
    (e) none of the above

Section 12.5 *Electron Dot Formulas of Polyatomic Ions*

11. Draw the electron dot formula for the carbonate ion, CO_3^{2-}. How many pairs of nonbonding electrons are in a carbonate ion?
    (a) 3
    (b) 4
    (c) 8
    (d) 12
    (e) none of the above

12. What type of bonds are in a carbonate ion, CO_3^{2-}?
    (a) 1 single bond and 2 double bonds
    (b) 2 single bonds and 1 double bond
    (c) 2 single bonds and 1 triple bond
    (d) 3 single bonds
    (e) none of the above

Section 12.6 *Polar Covalent Bonds*

13. Which of the following is a general trend for the electronegativity of elements in the periodic table?
    (a) increases from left to right; increases from bottom to top
    (b) increases from left to right; decreases from bottom to top
    (c) decreases from left to right; increases from bottom to top
    (d) decreases from left to right; decreases from bottom to top
    (e) none of the above

14. Using the electronegativity trends in the periodic table, predict which of the following molecules contain polar covalent bonds.
    (a) hydrogen chloride, HCl
    (b) hydrogen fluoride, HF
    (c) carbon monoxide, CO
    (d) nitrogen monoxide, NO
    (e) all of the above

15. Given the electronegativity values of I (2.5) and Br (2.8), illustrate the bond polarity in an iodine monobromide molecule, IBr, using delta notation.
    (a) $(\delta+)$ I—Br $(\delta+)$
    (b) $(\delta+)$ I— Br $(\delta-)$
    (c) $(\delta-)$ I— Br $(\delta+)$
    (d) $(\delta-)$ I— Br $(\delta-)$
    (e) none of the above

16. Given the electronegativity values for I (2.5) and Br (2.8), calculate the bond polarity in an iodine monobromide molecule, IBr.
    (a) –0.3
    (b) 0.3
    (c) 2.7
    (d) 5.3
    (e) –5.3

Section 12.7 *Nonpolar Covalent Bonds*

17. If the electronegativity values for H, N, O, and P are 2.1, 3.0, 3.5, and 2.1, respectively, which of the following is a nonpolar covalent molecule?
    (a) water, H_2O
    (b) laughing gas, N_2O
    (c) phosphine, PH_3
    (d) ammonia, NH_3
    (e) all of the above

18. Which of the following elements occurs naturally as a nonpolar diatomic molecule?
    (a) H_2
    (b) N_2
    (c) O_2
    (d) F_2
    (e) all of the above

Section 12.8 *Coordinate Covalent Bonds*

19. Which of the following molecules contains a coordinate covalent bond?
    (a) HClO
    (b) $HBrO_2$
    (c) HIO_3
    (d) $HClO_4$
    (e) all of the above

Section 12.9 *Hydrogen Bonds*

20. Which of the following describes the attraction between two H_2O molecules?
    (a) coordinate covalent bond
    (b) hydrogen bond
    (c) nonpolar covalent bond
    (d) polar covalent bond
    (e) none of the above

21. Which of the following is true of a hydrogen bond?
    (a) The bond is between a H atom and an O, N, or F atom.
    (b) The bond energy is greater than a covalent bond.
    (c) The bond length is shorter than a covalent bond.
    (d) all of the above
    (e) none of the above

Section 12.10 *Shapes of Molecules*

22. What is the electron pair geometry for a hydrogen selenide, H_2Se, molecule?
    (a) bent
    (b) linear
    (c) tetrahedral
    (d) trigonal planar
    (e) trigonal pyramidal

23. What is the molecular shape of a hydrogen selenide, H_2Se, molecule?
    (a) bent
    (b) linear
    (c) tetrahedral
    (d) trigonal planar
    (e) trigonal pyramidal

24. What is the electron pair geometry for an arsine, AsH_3, molecule?
    (a) bent
    (b) linear
    (c) tetrahedral
    (d) trigonal planar
    (e) trigonal pyramidal

25. What is the molecular shape of an arsine, AsH_3, molecule?
    (a) bent
    (b) linear
    (c) tetrahedral
    (d) trigonal planar
    (e) trigonal pyramidal

26. What is the electron pair geometry for a silicate, SiO_3^{2-}, polyatomic ion?
    (a) bent
    (b) linear
    (c) tetrahedral
    (d) trigonal planar
    (e) trigonal pyramidal

27. What is the molecular shape of a silicate, SiO_3^{2-}, polyatomic ion?
    (a) bent
    (b) linear
    (c) tetrahedral
    (d) trigonal planar
    (e) trigonal pyramidal

Electron Dot Formulas for Molecules and Polyatomic Ions

28. Calculate the total number of valence electrons for each of the following molecules and draw the electron dot and structural formulas.

   (a) PI_3

   (b) HOCN

29. Calculate the total number of valence electrons for each of the following polyatomic ions and draw the electron dot and structural formulas.

   (a) SO_3^{2-}

   (b) BO_3^{3-}

Across

1. a chloride ion but *not* a chlorite ion
4. a bond containing two electrons
7. a representative particle of an ionic compound
9. a bond containing four electrons
11. the attraction between two atoms
12. a bond containing six electrons
17. the ability to attract shared electrons
21. the rule involving eight valence electrons
23. the molecular shape of water
24. the electron pair geometry of water
25. a formula that shows valence electrons
26. valence electrons that are not shared
27. formed by two atoms bonded to a central atom
28. required to break a covalent bond

Down

1. a particle held together by covalent bonds
2. a bond resulting from the attraction of ions
3. a theory that explains molecular shape
5. a covalent bond with electrons from one atom
6. distance between nuclei in a covalent bond
8. a nitrate ion but *not* a nitride ion
10. a formula that shows bonds using dashes
13. a covalent bond with unequal sharing
14. valence electrons that are shared
15. the notation that indicates bond polarity
16. an intermolecular bond
18. a bond resulting from sharing electrons
19. a two-atom molecule
20. a covalent bond with equal sharing
22. the electrons in *s* and *p* sublevels

Chapter 12 Answers to Self-Test

1. **b** 2. **a** 3. **e** 4. **c** 5. **b** 6. **d** 7. **a** 8. **b** 9. **a** 10. **c** 11. **c** 12. **b** 13. **a** 14. **e** 15. **b** 16. **b** 17. **c** 18. **e** 19. **e** 20. **b** 21. **a** 22. **c** 23. **a** 24. **c** 25. **e** 26. **d** 27. **d**

Electron Dot Formulas for Molecules and Polyatomic Ions

28.

	Molecule	Valence Electrons	Electron Dot	Structural
(a)	**PI_3**	$5 + 3(7) = 26\ e^-$	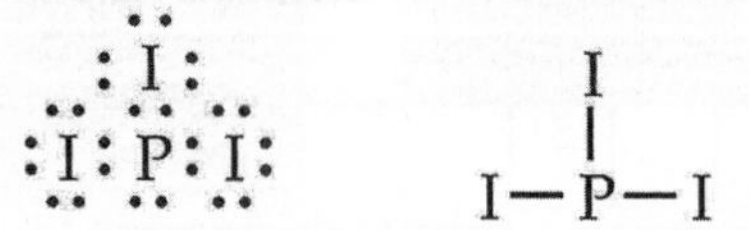	
(b)	H**OC**N	$1 + 6 + 4 + 5 = 16\ e^-$	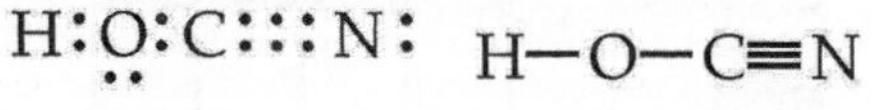	

29.

	Ion	Valence Electrons	Electron Dot	Structural
(a)	**SO_3^{2-}**	$4(6) + 2 = 26\ e^-$	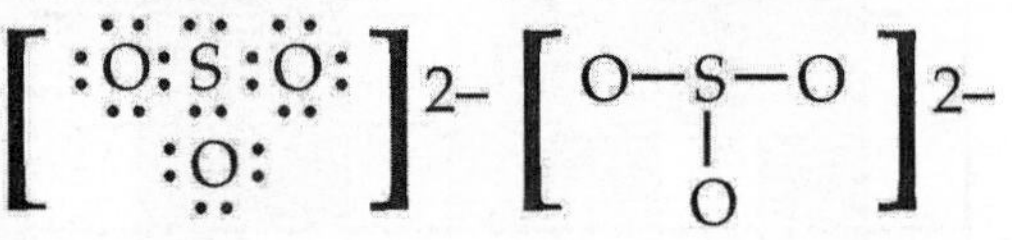	
(b)	**BO_3^{3-}**	$3 + 3(6) + 3 = 24\ e^-$	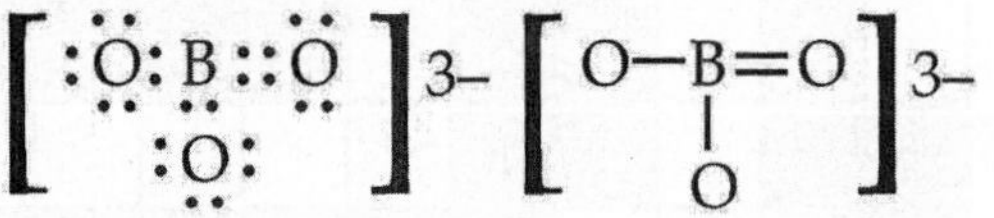	

Chapter 12 Answers to Crossword Puzzle

Across

1. monoatomic
4. single
7. unit
9. double
11. bond
12. triple
17. electronegativity
21. octet
23. bent
24. tetrahedral
25. dot
26. nonbonding
27. angle
28. energy

Down

1. molecule
2. ionic
3. VSEPR
5. coordinate
6. length
8. polyatomic
10. structural
13. polar
14. bonding
15. delta
16. hydrogen
18. covalent
19. diatomic
20. nonpolar
22. valence

Liquids and Solids

Section 13.1 *Properties of Liquids*

1. Which of the following are observed general properties of liquids?
    (a) liquids have a variable shape and fixed volume
    (b) liquids compress and expand significantly
    (c) liquids are much less dense than gases
    (d) liquids that are insoluble mix homogeneously
    (e) none of the above

2. Predict the physical state of argon at –190 °C (Mp = –189 °C, Bp = –186 °C) and normal atmospheric pressure.
    (a) gas
    (b) liquid
    (c) solid
    (d) solid and liquid
    (e) none of the above

Section 13.2 *The Intermolecular Bond Concept*

3. What is the strongest intermolecular force in a liquid containing polar molecules?
    (a) covalent bonds
    (b) ionic bonds
    (c) permanent dipole forces
    (d) temporary dipole forces
    (e) none of the above

4. What is the strongest intermolecular force in a liquid containing nonpolar molecules?
    (a) covalent bonds
    (b) dipole forces
    (c) dispersion forces
    (d) hydrogen bonds
    (e) none of the above

5. What is the strongest intermolecular force in a liquid having molecules with H–N bonds?
   (a) dipole forces
   (b) dispersion forces
   (c) hydrogen bonds
   (d) covalent bonds
   (e) none of the above

6. Given the following liquids have similar molar masses, which has the strongest molecular attraction?
   (a) $CH_3–CO–O–H$
   (b) $CH_3–CH_2–O–CH_3$
   (c) $CH_3–CH_2–S–CH_3$
   (d) $CH_3–CH_2–CH_2–Cl$
   (e) $CH_3–CH_2–CH_2–CH_2–CH_3$

Section 13.3 *Vapor Pressure, Boiling Point, Viscosity, Surface Tension*

7. Consider the following liquids with similar molar masses. Which liquid has the strongest attraction between molecules based only on boiling point data?
   (a) Liquid X   (Bp = 50 °C @ 760 mm Hg)
   (b) Liquid Y   (Bp = 0 °C @ 760 mm Hg)
   (c) Liquid Z   (Bp = –50 °C @ 760 mm Hg)
   (d) all liquids have the same intermolecular attraction
   (e) insufficient data to predict

8. Consider the following liquids with similar molar masses. Which liquid has the strongest attraction between molecules based on vapor pressure data?
   (a) Liquid X   (vapor pressure = 50 mm Hg @ 20 °C)
   (b) Liquid Y   (vapor pressure = 100 mm Hg @ 20 °C)
   (c) Liquid Z   (vapor pressure = 150 mm Hg @ 20 °C)
   (d) all liquids have the same intermolecular attraction
   (e) insufficient data to predict

9. Consider the following liquids with similar molar masses. Which liquid has the strongest attraction between molecules based on viscosity data?
   (a) Liquid X   (viscosity = 0.50 centipoise @ 20 °C)
   (b) Liquid Y   (viscosity = 0.75 centipoise @ 20 °C)
   (c) Liquid Z   (viscosity = 1.25 centipoise @ 20 °C)
   (d) all liquids have the same intermolecular attraction
   (e) insufficient data to predict

10. Consider the following liquids with similar molar masses. Which liquid has the strongest attraction between molecules based on surface tension data?
    (a) Liquid X   (surface tension = 15 dynes/cm @ 20 °C)
    (b) Liquid Y   (surface tension = 25 dynes/cm @ 20 °C)
    (c) Liquid Z   (surface tension = 50 dynes/cm @ 20 °C)
    (d) all liquids have the same intermolecular attraction
    (e) insufficient data to predict

Section 13.4 *Properties of Solids*

11. Which of the following properties is *not* a general characteristic of solids?
    (a) crystalline and noncrystalline structures
    (b) do not compress or expand significantly
    (c) heterogeneous solids mix by diffusion
    (d) rigid shape and fixed volume
    (e) usually more dense than corresponding liquids

12. Predict the physical state of argon at -180 °C (Mp $= -189$ °C, Bp $= -186$ °C) and normal atmospheric pressure.
    (a) gas
    (b) liquid
    (c) solid
    (d) solid and liquid
    (e) liquid and gas

Section 13.5 *Crystalline Solids*

13. Which of the following is an example of an ionic crystalline solid?
    (a) calcite, $CaCO_3$
    (b) fluorite, CaF_2
    (c) iron pyrite, FeS_2
    (d) silver chloride, $AgCl$
    (e) all of the above

14. Which of the following is *not* an example of a molecular crystalline solid?
    (a) dry ice, CO_2
    (b) rose gold, Au/Cu
    (c) sucrose, $C_{12}H_{22}O_{11}$
    (d) sulfur, S_8
    (e) urea, $CO(NH_2)_2$

15. Which of the following is an example of a metallic crystalline solid?
    (a) diamond, C
    (b) halite, $NaCl$
    (c) iodine, I_2
    (d) phosphorus, P_4
    (e) solder, Pb/Sn

Section 13.6 *Changes of Physical State*

16. Calculate the heat required to convert 10.0 g of ice at −10.0 °C to water at 20 °C. The specific heat of ice is 0.50 cal/(g × °C); the heat of fusion for ice is 80.0 cal/g; and the specific heat of water is 1.00 cal/(g × °C).
    (a) 105 cal
    (b) 825 cal
    (c) 1050 cal
    (d) 1200 cal
    (e) 2800 cal

17. Calculate the heat released when 25.0 g of steam at 100.0 °C cools to water at 0 °C. The heat of condensation for steam is 540.0 cal/g; and the specific heat of water is 1.00 cal/(g × °C).
    (a) 4500 cal
    (b) 13,500 cal
    (c) 15,500 cal
    (d) 16,000 cal
    (e) 18,000 cal

Section 13.7 *Structure of Water*

18. What is the number of nonbonding electron pairs in a water molecule?
    (a) 0   (b) 1   (c) 2   (d) 3   (e) 4

19. What is the experimentally observed bond angle in a water molecule?
    (a) 90°
    (b) 104.5°
    (c) 109.5°
    (d) 120°
    (e) 180°

20. Which of the following illustrates the bond polarity in a water molecule?
    (a) (δ−) O—H (δ+)
    (b) (δ−) O—H (δ−)
    (c) (δ+) O—H (δ+)
    (d) (δ+) O—H (δ−)
    (e) (δ)   O—H (δ)

Section 13.8 *Physical Properties of Water*

21. Which of the following accounts for the unusually high heats of fusion and vaporization for water?
    (a) the hydrogen bonds in water
    (b) the molar mass of water
    (c) the specific heat of water
    (d) the surface tension in water
    (e) the viscosity of water

22. Which of the following physical properties has a higher value for water (H_2O) than the corresponding value for heavy water, (D_2O)?
    (a) density
    (b) melting point
    (c) heat of vaporization
    (d) molar mass
    (e) none of the above

Section 13.9 *Chemical Properties of Water*

23. What are the products from the complete combustion of a hydrocarbon?
    (a) carbon and water
    (b) carbon monoxide and hydrogen
    (c) carbon monoxide and water
    (d) carbon dioxide and hydrogen
    (e) carbon dioxide and water

24. Complete the following chemical equation and indicate the product(s).

$$BaO(s) \ + \ H_2O(l) \ \rightarrow$$

    (a) $Ba(OH)_2$
    (b) $Ba(OH)_2 + H_2$
    (c) $Ba(OH)_2 + O_2$
    (d) $Ba(OH)_2 + O_3$
    (e) $Ba(OH)_2 + H_2O_2$

25. Complete the following chemical equation and indicate the product(s).

$$CO_2(g) \ + \ H_2O(l) \ \rightarrow$$

    (a) H_2CO_3
    (b) $HC_2H_3O_2$
    (c) CH_3OH
    (d) $CO + H_2O_2$
    (e) $CO + H_2 + O_2$

Section 13.10 *Hydrates*

26. What is the systematic name for $Ca(NO_3)_2 \cdot 4H_2O$?
    (a) calcium nitrate dihydrate
    (b) calcium nitrate tetrahydrate
    (c) calcium nitrite dihydrate
    (d) calcium nitrite tetrahydrate
    (e) none of the above

27. What is the percentage of water in borax, $Na_2B_4O_7 \cdot 10H_2O$?
    (a) 4.72%
    (b) 8.95%
    (c) 47.2%
    (d) 82.1%
    (e) 89.5%

Chemical Reactions Involving Water as a Reactant or Product

28. Complete and balance each of the following chemical reactions.

$$K(s) \quad + \quad H_2O(l) \quad \rightarrow$$

$$HNO_3(aq) \quad + \quad Zn(OH)_2(s) \quad \rightarrow$$

$$CH_3OH(g) \quad + \quad O_2(g) \quad \xrightarrow{spark}$$

$$KAl(SO_4)_2 \cdot 12H_2O(s) \quad \xrightarrow{\Delta}$$

Determing the Empirical Formula of a Hydrate

29. An unknown hydrate of chromium(III) acetate, $Cr(C_2H_3O_2)_3 \cdot XH_2O$, is heated to give 7.29% water. Calculate the water of crystallization (X) for the salt and state the formula and name of the hydrate.

Chapter 13 Key Terms

Across

4. the heat required to convert a liquid to a gas
8. a crystalline solid composed of ions
9. the resolution of two or more dipoles
10. a crystalline solid composed of molecules
15. the tendency of a liquid to form drops (2 words)
17. a substance containing attached water molecules
20. water containing sodium ions and various anions
23. reaction by passing electricity through a solution
24. the heat required to convert a solid to a liquid
25. water containing a variety of cations and anions

Down

1. temperature when vapor pressure is 1 atm
2. attractive force from temporary dipoles
3. pressure when condensation equals evaporation
5. an oxide that gives an acidic solution
6. a crystalline solid composed of metal atoms
7. arrangement of two atoms bonded to a central atom
11. an intermolecular bond between H and O
12. water of crystallization
13. the resistance of a liquid to flow
14. a solid with particles that repeat in a pattern
16. a compound that does not contain water
18. attractive force from permanent dipoles
19. demineralized water
21. an oxide that gives a basic solution
22. a molecule of water containing deuterium

1. a   2. c   3. c   4. c   5. c   6. a   7. a   8. a   9. c   10. c   11. c   12. a   13. e   14. b   15. e
16. c   17. d   18. c   19. b   20. a   21. a   22. e   23. e   24. a   25. a   26. b   27. c

Chemical Reactions Involving Water as a Reactant or Product

28.

$$2\,K(s) \; + \; 2\,H_2O(l) \quad \rightarrow \quad 2\,KOH(aq) \; + \; H_2(g)$$

$$2\,HNO_3(aq) \; + \; Zn(OH)_2(s) \quad \rightarrow \quad Zn(NO_3)_2(aq) \; + \; 2\,H_2O(l)$$

$$2\,CH_3OH(g) \; + \; 3\,O_2(g) \quad \rightarrow \quad 2\,CO_2(g) \; + \; 4\,H_2O(g)$$

$$KAl(SO_4)_2 \cdot 12H_2O(s) \quad \rightarrow \quad KAl(SO_4)_2(s) \; + \; 12\,H_2O(g)$$

Determing the Empirical Formula of a Hydrate

29.

$$Cr(C_2H_3O_2)_3 \cdot XH_2O(s) \quad \overset{\Delta}{\rightarrow} \quad Cr(C_2H_3O_2)_3(s) + XH_2O(g)$$

$$7.29 \; \cancel{g\,H_2O} \; \times \; \frac{1 \; mol \; H_2O}{18.02 \; \cancel{g\,H_2O}} \quad = \quad 0.405 \; mol \; H_2O$$

$$92.71 \; \cancel{g\,Cr(C_2H_3O_2)_3} \; \times \; \frac{1 \; mol \; Cr(C_2H_3O_2)_3}{229.15 \; \cancel{g\,Cr(C_2H_3O_2)_3}} \quad = \quad 0.4046 \; mol \; Cr(C_2H_3O_2)_3$$

$$\text{Water of Crystallization:} \; \frac{0.405}{0.4046} \quad = \quad 1.00 \approx 1$$

Formula: $Cr(C_2H_3O_2)_3 \cdot H_2O$ **Name:** chromium(III) acetate monohydrate

Chapter 13 Answers to Crossword Puzzle

Across
4. vaporization
8. ionic
9. net
10. molecular
15. surface tension
17. hydrate
20. soft
23. electrolysis
24. fusion
25. hard

Down
1. Bp
2. dispersion
3. vapor
5. nonmetal
6. metallic
7. angle
11. hydrogen
12. hydration
13. viscosity
14. crystalline
16. anhydrous
18. dipole
19. deionized
21. metal
22. heavy

Solutions

Section 14.1 *Gases in Solution*

1. Carbonated drinks are injected with carbon dioxide gas. Under what conditions does carbon dioxide gas have the greatest solubility?
   (a) low temperature, low pressure
   (b) low temperature, high pressure
   (c) low temperature, pressure is not a factor
   (d) high pressure, temperature is not a factor
   (e) high temperature, high pressure

2. If the solubility of nitrogen gas in blood is 1.90 mL per 100 mL at one atmosphere, what is the solubility of nitrogen gas in a scuba diver's blood at a depth of 155 feet where the pressure is 5.50 atmospheres?
   (a) 0.190 mL/100 mL
   (b) 0.345 mL/100 mL
   (c) 1.90 mL/100 mL
   (d) 3.80 mL/100 mL
   (e) 10.5 mL/100 mL

Section 14.2 *Liquids in Solution*

3. Which of the following illustrates the *like dissolves like* rule for two liquids?
   (a) a polar solvent is insoluble in a polar solvent
   (b) a polar solvent is soluble in a nonpolar solvent
   (c) a nonpolar solvent is soluble in a nonpolar solvent
   (d) all of the above
   (e) none of the above

4. Apply the *like dissolves like* rule to predict which of the following liquids is miscible with water.
   (a) chloroform, $CHCl_3$
   (b) glycerin, $C_3H_5(OH)_3$
   (c) toluene, $C_6H_5CH_3$
   (d) all of the above
   (e) none of the above

Section 14.3 Solids in Solution

5. Which of the following illustrates the *like dissolves like* rule for a solid solute in a liquid solvent?
   (a) an ionic compound is insoluble in a polar solvent
   (b) a polar compound is insoluble in a nonpolar solvent
   (c) a nonpolar compound is insoluble in a nonpolar solvent
   (d) all of the above
   (e) none of the above

6. Apply the *like dissolves like* rule to predict which of the following vitamins is insoluble in water.
   (a) thiamine, vitamin B_1, $C_{12}H_{18}Cl_2N_4OS$
   (b) riboflavin, vitamin B_2, $C_{17}H_{20}N_4O_6$
   (c) ascorbic acid, vitamin C, $C_6H_8O_6$
   (d) calciferol, vitamin D, $C_{27}H_{44}O$
   (e) all of the above

7. Apply the *like dissolves like* rule to predict which of the following is insoluble in benzene, C_6H_6.
   (a) benzopyrene (a compound in charcoal), $C_{20}H_{12}$
   (b) DDT (an insecticide), $C_{14}H_9Cl_5$
   (c) glycine (an amino acid), $CH_2(NH_2)COOH$
   (d) naphthalene (mothballs), $C_{10}H_8$
   (e) paradichlorobenzene (mothballs), $C_6H_4Cl_2$

Section 14.4 The Dissolving Process

8. When fructose, $C_6H_{12}O_6$, dissolves in water, which of the following is formed in solution?
   (a) acetic acid, $HC_2H_3O_2$
   (b) carbonic acid, H_2CO_3
   (c) hydrated clusters of CO_2 molecules
   (d) hydrated clusters of $C_6H_{12}O_6$ molecules
   (e) hydrated clusters of hydrogen and oxygen molecules

9. When solid potassium fluoride dissolves in water, which of the following is one of the aqueous ions formed?

   (a) K^+ --- $\underset{\underset{H}{|}}{H-O}$ (b) K^+ --- $\underset{\underset{H}{|}}{O-H}$ (c) F^- --- $\underset{\underset{H}{|}}{O-H}$

   (d) all of the above   (e) none of the above

Section 14.5 *Rate of Dissolving*

10. Which of the following increases the rate of dissolving for a solid solute in a solvent?
    (a) grinding the solute
    (b) heating the solution
    (c) stirring the solution
    (d) all of the above
    (e) none of the above

Section 14.6 *Solubility and Temperature*

11. What is the solubility of sucrose sugar, $C_{12}H_{22}O_{11}$, at 20 °C?
    (Refer to textbook Figure 14.5.)
    (a) 55 g/100 g water
    (b) 75 g/100 g water
    (c) 90 g/100 g water
    (d) 100 g/100 g water
    (e) 130 g/100 g water

12. What is the minimum temperature required to dissolve 140 g of sucrose sugar, $C_{12}H_{22}O_{11}$, in 100 g of water? *(Refer to textbook Figure 14.5.)*
    (a) 20 °C
    (b) 50 °C
    (c) 55 °C
    (d) 60 °C
    (e) 100 °C

Section 14.7 *Unsaturated, Saturated, and Supersaturated Solutions*

13. How concentrated is a solution containing 100 g of $NaC_2H_3O_2$ in 100 g water at 75 °C? *(Refer to textbook Figure 14.6.)*
    (a) unsaturated
    (b) saturated
    (c) supersaturated
    (d) superunsaturated
    (e) none of the above

14. How concentrated is a solution containing 100 g of $NaC_2H_3O_2$ in 100 g water at 25 °C? *(Refer to textbook Figure 14.6.)*
    (a) unsaturated
    (b) saturated
    (c) supersaturated
    (d) superunsaturated
    (e) none of the above

Section 14.8 *Mass/Mass Percent Concentration*

15. Which of the following is *not* a unit factor related to a 5.00% aqueous solution of sodium hydroxide, NaOH?

    (a) $\dfrac{5.00 \text{ g NaOH}}{95.00 \text{ g solution}}$ (b) $\dfrac{95.00 \text{ g water}}{5.00 \text{ g NaOH}}$

    (c) $\dfrac{5.00 \text{ g NaOH}}{95.00 \text{ g water}}$ (d) $\dfrac{100 \text{ g solution}}{5.00 \text{ g NaOH}}$

    (e) $\dfrac{95.00 \text{ g water}}{100 \text{ g solution}}$

16. What is the mass of a 10.0% blood plasma solution that contains 2.50 g of dissolved solute?
    (a) 0.250 g
    (b) 0.278 g
    (c) 22.5 g
    (d) 25.0 g
    (e) 250 g

17. What is the mass of glucose, $C_6H_{12}O_6$, solute in 10.0 g of a 5.00% sugar solution?
    (a) 0.180 g
    (b) 0.500 g
    (c) 0.900 g
    (d) 9.50 g
    (e) 10.0 g

18. What is the mass of water needed to prepare 5.00 kg of a 40.0% antifreeze solution?
    (a) 2.00 kg
    (b) 3.00 kg
    (c) 3.33 kg
    (d) 12.5 kg
    (e) 200 kg

19. An hour after a meal, a patient's blood glucose level is 120 mg/dL. Express this glucose level as mass/mass percent concentration. (Assume the density of blood is 1.0 g/mL.)
    (a) 12%
    (b) 1.2%
    (c) 0.12%
    (d) 0.012%
    (e) 0.0012%

Section 14.9 *Molar Concentration*

20. Which of the following is *not* a unit factor related to a 0.500 M NaOH solution?

    (a) $\dfrac{0.500 \text{ mol NaOH}}{1 \text{ L solution}}$

    (b) $\dfrac{1 \text{ L solution}}{0.500 \text{ mol NaOH}}$

    (c) $\dfrac{0.500 \text{ mol NaOH}}{1000 \text{ mL solution}}$

    (d) $\dfrac{1000 \text{ mL solution}}{0.500 \text{ mol NaOH}}$

    (e) $\dfrac{0.500 \text{ mol NaOH}}{1000 \text{ g solution}}$

21. What is the mass of barium hydroxide (171.35 g/mol) dissolved in 250.0 mL of 0.200 M Ba(OH)$_2$ solution?
    (a) 8.57 g
    (b) 17.1 g
    (c) 85.7 g
    (d) 171 g
    (e) 857 g

22. What volume of 6.00 M sulfuric acid contains 9.80 g of H$_2$SO$_4$ (98.03 g/mol)?
    (a) 0.600 mL
    (b) 16.7 mL
    (c) 60.0 mL
    (d) 167 mL
    (e) 1670 mL

Section 14.10 *Dilution of a Solution*

23. Dilute sulfuric acid is a 6.0 M solution. What is the molarity of a H$_2$SO$_4$ solution prepared by diluting 50.0 mL of dilute acid to a total volume of 1.00 L?
    (a) 0.10 M
    (b) 0.12 M
    (c) 0.30 M
    (d) 0.50 M
    (e) 3.33 M

Section 14.11 *Solution Stoichiometry*

24. Given that 25.0 mL of 0.100 M barium chloride solution reacts completely with 22.5 mL of silver nitrate solution, what is the molarity of the AgNO$_3$ solution?

$$\text{BaCl}_2(aq) \quad + \quad 2\,\text{AgNO}_3(aq) \quad \rightarrow \quad 2\,\text{AgCl}(s) \quad + \quad \text{Ba(NO}_3)_2(aq)$$

    (a) 0.0450 M
    (b) 0.0556 M
    (c) 0.111 M
    (d) 0.222 M
    (e) 0.180 M

Mass/Mass Percent and Molar Concentrations

25. A 50.0-mL sample of solution has a mass of 53.810 g and contains 5.379 g of solid KI. Calculate the mass/mass percent concentration of the KI solution.

26. Given the data in the preceding problem, calculate the molar concentration of the KI solution.

Chapter 14 Key Terms

Across

2. a solvent composed of polar molecules
3. liquids that dissolve in one another
5. the lesser component of a solution
7. a solvent composed of nonpolar molecules
8. stated gas solubility is proportional to pressure
11. mol solute/L of solution concentration
12. a mixture of 1–100 nm dispersed particles
16. a solution with more than maximum solute
19. a colloidal solution
20. the overall direction of negative charge (2 words)
21. "like _____ like" rule

Down

1. the greater component of a solution
4. relates quantities according to a balanced equation
6. a solution with less than maximum solute
9. liquids that do not dissolve in one another
10. a solute dissolved in a solvent
13. a region with partial negative and positive charge
14. solvent molecules surrounding a solute particle
15. a solution with the maximum solute
16. amount of solute that will dissolve in a solvent
17. g solute/100 g of solution concentration
18. type of light scattering by colloidal particles

1. **b** 2. **e** 3. **c** 4. **b** 5. **b** 6. **d** 7. **c** 8. **d** 9. **b** 10. **d** 11. **d** 12. **c** 13. **a** 14. **c** 15. **a**
16. **d** 17. **b** 18. **b** 19. **c** 20. **e** 21. **a** 22. **b** 23. **c** 24. **d**

Mass/Mass Percent Concentration and Molar Concentrations

25. $$\frac{5.379 \text{ g KI}}{53.810 \text{ g soln}} \times 100\% \quad = \quad \text{m/m \% KI}$$

$$= \quad 9.996\% \text{ KI}$$

26. $$\frac{5.379 \text{ g KI}}{50.0 \text{ mL soln}} \times \frac{1 \text{ mol KI}}{166.00 \text{ g KI}} \times \frac{1000 \text{ mL soln}}{1 \text{ L soln}} \quad = \quad \text{mol KI/L soln}$$

$$= \quad 0.648 \, M \text{ KI}$$

Chapter 14 Answers to Crossword Puzzle

Across
2. polar
3. miscible
5. solute
7. nonpolar
8. Henry
11. molarity
12. colloid
16. supersaturated
19. dispersion
20. net dipole
21. dissolves

Down
1. solvent
4. stoichiometry
6. unsaturated
9. immiscible
10. solution
13. dipole
14. cage
15. saturated
16. solubility
17. percent
18. Tyndall

Acids and Bases

Section 15.1 *Properties of Acids and Bases*

1. Which of the following is a general property of an acidic solution?
   (a) tastes sour
   (b) turns litmus paper red
   (c) pH less than 7
   (d) reacts with a basic solution to give a salt and water
   (e) all of the above

2. If a solution has a pH of 11, which of the following describes the solution?
   (a) strongly acidic
   (b) weakly acidic
   (c) neutral
   (d) weakly basic
   (e) strongly basic

Section 15.2 *Arrhenius Acids and Bases*

3. Given the ionization, which of the following is a strong Arrhenius acid?
   (a) $HClO_2(aq)$ ($\sim 1\%$ ionized)
   (b) $H_2SO_3(aq)$ ($\sim 1\%$ ionized)
   (c) $H_3PO_4(aq)$ ($\sim 1\%$ ionized)
   (d) all of the above
   (e) none of the above

4. What acid and base are neutralized to give the salt potassium chloride?
   (a) $HCl(aq)$ and $KOH(aq)$
   (b) $HCl(aq)$ and $KCl(aq)$
   (c) $HOH(aq)$ and $KCl(aq)$
   (d) $HOH(aq)$ and $KOH(aq)$
   (e) none of the above

Section 15.3 Brønsted–Lowry Acids and Bases

5. In the following reaction, which reactant is acting as a Brønsted–Lowry acid?

$$Na_2HPO_4(aq) \; + \; H_2CO_3(aq) \quad \rightarrow \quad NaHCO_3(aq) \; + \; NaH_2PO_4(aq)$$

(a) Na_2HPO_4
(b) H_2CO_3
(c) $NaHCO_3$
(d) NaH_2PO_4
(e) H_2O

6. In the following reaction, which reactant is acting as a Brønsted–Lowry base?

$$NaHCO_3(aq) \; + \; NaH_2PO_4(aq) \quad \rightarrow \quad Na_2HPO_4(aq) \; + \; H_2CO_3(aq)$$

(a) $NaHCO_3$
(b) NaH_2PO_4
(c) Na_2HPO_4
(d) H_2CO_3
(e) H_2O

Section 15.4 Acid–Base Indicators

7. Which of the following acid–base indicators is red in an acidic solution and yellow in a basic solution?
(a) methyl red
(b) bromthymol blue
(c) phenolphthalein
(d) all of the above
(e) none of the above

8. Which of the following acid–base indicators is yellow in an acidic solution and blue in a basic solution?
(a) methyl red
(b) bromthymol blue
(c) phenolphthalein
(d) all of the above
(e) none of the above

9. Which of the following acid–base indicators is colorless in an acidic solution and pink in a basic solution?
(a) methyl red
(b) bromthymol blue
(c) phenolphthalein
(d) all of the above
(e) none of the above

Section 15.5 *Acid–Base Titrations*

10. If 10.0 mL of 0.500 M $HC_2H_3O_2$ is titrated with 0.250 M NaOH, what volume of sodium hydroxide solution is required to neutralize the acid?

$$HC_2H_3O_2(aq) + NaOH(aq) \rightarrow NaC_2H_3O_2(aq) + H_2O(l)$$

(a) 5.00 mL
(b) 10.0 mL
(c) 20.0 mL
(d) 40.0 mL
(e) 80.0 mL

11. If a 0.500 M $HC_2H_3O_2$ solution has a density of 1.00 g/mL, what is the mass percent concentration of the acetic acid solution?
(a) 2.00%
(b) 3.00%
(c) 5.00%
(d) 10.0%
(e) 50.0%

Section 15.6 *Acid–Base Standardization*

12. What is the molarity of a potassium hydroxide solution if 25.00 mL of KOH is required to neutralize 2.042 g of $KHC_8H_4O_4$ (204.23 g/mol)?

$$KHC_8H_4O_4(aq) + KOH(aq) \rightarrow K_2C_8H_4O_4(aq) + H_2O(l)$$

(a) 0.200 M
(b) 0.250 M
(c) 0.400 M
(d) 0.500 M
(e) 0.800 M

13. Vitamin C has the chemical name ascorbic acid and can be abbreviated HAsc. If 32.00 mL of 0.400 M NaOH neutralizes 2.240 g of ascorbic acid, what is the molar mass of vitamin C?

$$HAsc(aq) + NaOH(aq) \rightarrow NaAsc(aq) + H_2O(l)$$

(a) 28.0 g/mol
(b) 87.4 g/mol
(c) 175 g/mol
(d) 180 g/mol
(e) 350 g/mol

Section 15.7 *Ionization of Water*

14. Which of the following is the ionization constant expression for water?
    (a) $K_w = [H_2O] [H_2O]$
    (b) $K_w = [H_2O] [H_3O^+] / [OH^-]$
    (c) $K_w = [H_2O] [OH^-] / [H_3O^+]$
    (d) $K_w = [H_3O^+] [OH^-]$
    (e) $K_w = [H_3O^+] [OH^-] / [H_2O]$

15. Given an aqueous solution in which the $[H^+] = 5.0 \times 10^{-5}$ M, what is the molar hydroxide ion concentration?
    (a) $[OH^-] = 2.0 \times 10^{-8}$ M
    (b) $[OH^-] = 2.0 \times 10^{-9}$ M
    (c) $[OH^-] = 2.0 \times 10^{-10}$ M
    (d) $[OH^-] = 5.0 \times 10^{-9}$ M
    (e) $[OH^-] = 5.0 \times 10^{-10}$ M

16. Which of the following is true if the hydrogen ion concentration of an aqueous solution increases?
    (a) the pH increases
    (b) the K_w increases
    (c) the K_w decreases
    (d) the hydroxide ion concentration increases
    (e) the hydroxide ion concentration decreases

Section 15.8 *The pH Concept*

17. What is the pH of a solution with a 0.001 molar hydrogen ion concentration?
    (a) 1
    (b) 2
    (c) 3
    (d) 4
    (e) 10

18. What is the molar hydrogen ion concentration of stomach acid that registers a pH of 2 on a strip of pH paper?
    (a) 0.02 M
    (b) 0.01 M
    (c) 0.2 M
    (d) 0.1 M
    (e) 2 M

Section 15.9 *pH Calculations*

19. What is the pH of a sample solution if the $[H^+] = 0.000\ 065\ M$?
    (a) 0.81
    (b) 4.19
    (c) 4.81
    (d) 5.19
    (e) 7.81

20. What is the $[H^+]$ of a sample solution if the pH is 10.30?
    (a) 1.01 M
    (b) 0.000 000 000 20 M
    (c) 0.000 000 000 50 M
    (d) 0.000 000 000 020 M
    (e) 0.000 000 000 050 M

Section 15.10 *Strong and Weak Electrolytes*

21. If a light bulb in a conductivity apparatus glows brightly when testing a solution, which of the following is true?
    (a) The solution is highly ionized.
    (b) The solution is slightly ionized.
    (c) The solution is nonionized.
    (d) The solution is highly reactive.
    (e) The solution is slightly reactive.

22. Which of the following aqueous solutions are strong electrolytes?
    (a) strong acids
    (b) strong bases
    (c) soluble salts
    (d) all of the above
    (e) none of the above

23. Sodium chloride is a soluble salt and silver chloride is a slightly soluble salt. Which of the following correctly portrays aqueous solutions of the two salts?
    (a) NaCl(*aq*) and AgCl(*s*)
    (b) Na⁺(*aq*) + Cl⁻(*aq*) and AgCl(*s*)
    (c) NaCl(*aq*) and Ag⁺(*aq*) + Cl⁻(*aq*)
    (d) Na⁺(*aq*) + Cl⁻(*aq*) and Ag⁺(*aq*) + Cl⁻(*aq*)
    (e) none of the above

Section 15.11 *Net Ionic Equations*

24. Write a balanced net ionic equation given the following total ionic equation:

$H^+(aq) + NO_3^-(aq) + NH_4OH(aq) \rightarrow NH_4^+(aq) + NO_3^-(aq) + H_2O(l)$

(a) $HNO_3(aq) + NH_4OH(aq) \rightarrow NH_4NO_3(aq) + H_2O(l)$
(b) $H^+(aq) + OH^-(aq) \rightarrow H_2O(l)$
(c) $NO_3^-(aq) + NH_4^+(aq) \rightarrow NH_4NO_3(aq)$
(d) $H^+(aq) + NH_4OH(aq) \rightarrow NH_4^+(aq) + H_2O(l)$
(e) $HNO_3(aq) + OH^-(aq) \rightarrow NO_3^-(aq) + H_2O(l)$

Total and Net Ionic Equations

25. Write the total and net ionic equations for the following reaction.

$H_2SO_4(aq) + 2\,NaOH(aq) \rightarrow Na_2SO_4(aq) + 2\,H_2O(l)$

Acid–Base Titration

26. A 0.950-g sample of sodium carbonate is titrated with hydrochloric acid using bromcresol green indicator. Calculate the molar concentration of the acid if 38.10 mL of HCl are required to completely neutralize the sodium carbonate.

$HCl(aq) + Na_2CO_3(s) \rightarrow NaCl(aq) + H_2O(l) + CO_2(g)$

Chapter 15 Key Terms

Across

3. a polar compound dissolving in water
5. a Brønsted–Lowry base is a proton _____
8. ions that do not appear in the net ionic equation
10. a substance that can accept or donate a proton
12. an ionic equation showing all ionized substances
13. an ionic compound dissolving in water
17. a substance that releases hydroxide ions in water
19. an ionic equation with no spectator ions
20. a solution that resists changes in pH
21. an Arrhenius acid releases _____ ions
22. the hydrogen ion in aqueous solution

Down

1. a Brønsted–Lowry acid is a proton _____
2. an electrolyte that is a poor conductor of electricity
3. a substance that changes color with pH
4. the product of H^+ and OH^- concentrations
6. the molar H^+ expressed on an exponential scale
7. a product from a neutralization reaction
9. a solution whose concentration is known precisely
11. an Arrhenius base donates _____ ions
14. an electrolyte that is a good conductor
15. a substance that releases hydrogen ions in water
16. a procedure for delivering a solution with a buret
18. an indicator color change during a titration

1. e   2. d   3. e   4. a   5. b   6. a   7. a   8. b   9. c   10. c   11. b   12. c   13. c   14. d   15. c
16. e   17. c   18. b   19. b   20. e   21. a   22. d   23. b   24. d

Total Ionic and Net Ionic Equations

25. $$H_2SO_4(aq) + 2\,NaOH(aq) \rightarrow Na_2SO_4(aq) + 2\,H_2O(l)$$

$$2\,H^+(aq) + SO_4{}^{2-}(aq) + 2\,Na^+(aq) + 2\,OH^-(aq) \rightarrow 2\,Na^+(aq) + SO_4{}^{2-}(aq) + 2\,H_2O(l)$$

$$H^+(aq) + OH^-(aq) \rightarrow H_2O(l)$$

Acid–Base Titration

26. $$2\,HCl(aq) + Na_2CO_3(s) \rightarrow 2\,NaCl(aq) + H_2O(l) + CO_2(g)$$

$$0.950\ \cancel{g\ Na_2CO_3} \times \frac{1\ \cancel{mol\ Na_2CO_3}}{105.99\ \cancel{g\ Na_2CO_3}} \times \frac{2\ mol\ HCl}{1\ \cancel{mol\ Na_2CO_3}} = 0.0179\ mol\ HCl$$

$$\frac{0.0179\ mol\ HCl}{38.10\ \cancel{mL\ soln}} \times \frac{1000\ \cancel{mL\ soln}}{1\ L\ soln} = mol\ HCl/L\ soln$$

$$= 0.470\ M\ HCl$$

Across
- 3. ionization
- 5. acceptor
- 8. spectator
- 10. amphiprotic
- 12. total
- 13. dissociation
- 17. base
- 19. net
- 20. buffer
- 21. hydrogen
- 22. hydronium

Down
- 1. donor
- 2. weak
- 3. indicator
- 4. constant
- 6. pH
- 7. salt
- 9. standard
- 11. hydroxide
- 14. strong
- 15. acid
- 16. titration
- 18. endpoint

Chemical Equilibrium

Section 16.1 *Collision Theory*

1. Which of the following factors affects the rate of a chemical reaction?
   - (a) collision frequency
   - (b) collision energy
   - (c) collision orientation
   - (d) all of the above
   - (e) none of the above

2. Which of the following increases the collision frequency of molecules?
   - (a) increasing the temperature; adding a catalyst
   - (b) decreasing the temperature; adding a catalyst
   - (c) decreasing the concentration; adding a catalyst
   - (d) increasing the concentration; decreasing the temperature
   - (e) increasing the concentration; increasing the temperature

3. Which of the following increases the collision energy of molecules?
   - (a) increasing the temperature
   - (b) increasing the concentration
   - (c) increasing the collision frequency
   - (d) decreasing the volume of the container
   - (e) adding a catalyst

4. Which of the following factors increases the rate of a chemical reaction?
   - (a) decreasing the concentration of reactants
   - (b) decreasing the temperature
   - (c) adding a catalyst
   - (d) all of the above
   - (e) none of the above

5. Which of the following increases the amount of product from a reaction?
   - (a) adding a metal catalyst
   - (b) adding an acid catalyst
   - (c) using an UV light catalyst
   - (d) all of the above
   - (e) none of the above

Section 16.2 *Energy Profiles of Chemical Reactions*

6. If the E_{act} is lowered, which of the following is always true?
   (a) the reaction is exothermic
   (b) the reaction is endothermic
   (c) the reaction proceeds faster
   (d) the reaction proceeds slower
   (e) none of the above

7. If the heat of reaction is exothermic, which of the following is always true?
   (a) the reaction rate is fast
   (b) the reaction rate is slow
   (c) the energy of the reactants is less than the products
   (d) the energy of the reactants is greater than the products
   (e) none of the above

8. Which of the following lowers the ΔH for a chemical reaction?
   (a) increasing reactant concentration
   (b) increasing product concentration
   (c) adding a catalyst
   (d) all of the above
   (e) none of the above

Section 16.3 *The Chemical Equilibrium Concept*

9. Which of the following is true before a reaction reaches chemical equilibrium?
   (a) The amount of reactants is increasing.
   (b) The amount of products is increasing.
   (c) The amount of reactants and products are constant.
   (d) The amount of reactants and products are equal.
   (e) none of the above

10. Which of the following is true after a reaction reaches chemical equilibrium?
    (a) The amount of reactants is increasing.
    (b) The amount of products is increasing.
    (c) The amount of reactants and products are constant.
    (d) The amount of reactants and products are equal.
    (e) none of the above

11. Which of the following is true before a reaction reaches chemical equilibrium?
    (a) The rate of the forward reaction is increasing; the rate of the reverse reaction is decreasing.
    (b) The rate of the forward reaction is decreasing; the rate of the reverse reaction is increasing.
    (c) The rates of the forward and reverse reactions are increasing.
    (d) The rates of the forward and reverse reactions are decreasing.
    (e) The rates of the forward and reverse reactions are equal.

12. What is the equilibrium constant expression for: $3 A + B \leftrightarrow 2 C$?
    (a) $K_{eq} = [A][B] / [C]$
    (b) $K_{eq} = [A]^3[B] / [C]^2$
    (c) $K_{eq} = [C] / [A][B]$
    (d) $K_{eq} = [C]^2 / [A]^3[B]$
    (e) none of the above

13. What is the equilibrium constant expression for:

$$2 CO(g) \quad + \quad O_2(g) \quad \leftrightarrow \quad 2 CO_2(g)$$

    (a) $K_{eq} = [CO_2]^2 / [CO]^2[O_2]$
    (b) $K_{eq} = [CO_2]^2 / [CO]^2[O_2]^2$
    (c) $K_{eq} = [CO]^2[O_2] / [CO_2]^2$
    (d) $K_{eq} = [CO]^2[O_2]^2 / [CO_2]^2$
    (e) none of the above

14. What is the equilibrium constant expression for:

$$2 S(s) \quad + \quad 3 O_2(g) \quad \leftrightarrow \quad 2 SO_3(g)$$

    (a) $K_{eq} = [S][O_2]^3 / [SO_3]^2$
    (b) $K_{eq} = [S][O_2]^2 / [SO_3]^3$
    (c) $K_{eq} = [SO_3]^2 / [S][O_2]^3$
    (d) $K_{eq} = [SO_3]^3 / [S][O_2]^2$
    (e) $K_{eq} = [SO_3]^2 / [O_2]^3$

15. Oxygen and ozone gases are in chemical equilibrium in the upper atmosphere. Calculate K_{eq} for the reaction given the equilibrium concentrations of each gas at $20\,°C$: $[O_2] = 9.38 \times 10^{-3}$ and $[O_3] = 3.40 \times 10^{-15}$.

$$2 O_3(g) \quad \leftrightarrow \quad 3 O_2(g)$$

    (a) $K_{eq} = 4.11 \times 10^{-28}$
    (b) $K_{eq} = 1.40 \times 10^{-23}$
    (c) $K_{eq} = 3.62 \times 10^{-13}$
    (d) $K_{eq} = 2.43 \times 10^8$
    (e) $K_{eq} = 7.14 \times 10^{22}$

16. Which of the changes listed below will shift the equilibrium to the right for the following reversible reaction?

$$4 \, HCl(g) \; + \; O_2(g) \; + \; heat \quad \leftrightarrow \quad 2 \, Cl_2(g) \; + \; 2 \, H_2O(g)$$

(a) increase $[Cl_2]$
(b) decrease $[HCl]$
(c) decrease temperature
(d) increase pressure
(e) add a catalyst

17. Which of the changes listed below has no effect on the equilibrium for the following reversible reaction?

$$C(s) \; + \; O_2(g) \quad \leftrightarrow \quad CO_2(g) \; + \; heat$$

(a) increase $[CO_2]$
(b) increase $[O_2]$
(c) decrease $[O_2]$
(d) increase temperature
(e) increase pressure

Section 16.6 *Ionization Equilibrium Constant,* K_i

18. What is the equilibrium constant expression, K_i, for the weak acid H_2A?

$$H_2A(aq) \quad \leftrightarrow \quad H^+(aq) \; + \; HA^-(aq)$$

(a) $K_i = [H^+] \, [HA^-] \, / \, [H_2A]$
(b) $K_i = [H^+]^2 \, [HA^-] \, / \, [H_2A]$
(c) $K_i = [H^+]^2 \, [A^-] \, / \, [H_2A]$
(d) $K_i = [H_2A] \, / \, [H^+] \, [HA^-]$
(e) $K_i = [H_2A] \, / \, [H^+]^2 \, [A^-]$

19. If the hydrogen ion concentration of 0.100 M chloroacetic acid, $HC_2H_2O_2Cl$, is 0.012 M, what is the ionization constant for the acid?
(a) $K_i = 1.2 \times 10^{-1}$
(b) $K_i = 1.2 \times 10^{-3}$
(c) $K_i = 1.4 \times 10^{-3}$
(d) $K_i = 1.4 \times 10^{-4}$
(e) $K_i = 1.4 \times 10^{-5}$

Section 16.7 *Equilibria Shifts for Weak Acids and Bases*

20. Which of the changes listed below will shift the equilibrium to the right for the following reversible reaction in aqueous solution?

$$HC_2H_3O_2(aq) \quad \leftrightarrow \quad H^+(aq) \; + \; C_2H_3O_2^-(aq)$$

    (a)   decrease pH
    (b)   decrease $[HC_2H_3O_2]$
    (c)   add solid $NaC_2H_3O_2$
    (d)   add solid NaOH
    (e)   none of the above

21. Which of the changes listed below will have no effect on the equilibrium for the following reversible reaction in aqueous solution?

$$NH_4OH(aq) \quad \leftrightarrow \quad NH_4^+(aq) \; + \; OH^-(aq)$$

    (a)   add solid NH_4Cl
    (b)   add solid KCl
    (c)   add gaseous HCl
    (d)   add gaseous NH_3
    (e)   all of the above

Section 16.8 *Solubility Product Equilibrium Constant, K_{sp}*

22. What is the equilibrium constant expression, K_{sp}, for slightly soluble silver phosphate in aqueous solution?

$$Ag_3PO_4(s) \quad \leftrightarrow \quad 3\,Ag^+(aq) \; + \; PO_4^{3-}(aq)$$

    (a) $K_{sp} \; = \; [Ag^+]\,[PO_4^{3-}]$
    (b) $K_{sp} \; = \; [Ag^+]\,[PO_4^{3-}]^3$
    (c) $K_{sp} \; = \; [Ag^+]^3\,[PO_4^{3-}]$
    (d) $K_{sp} \; = \; [Ag^+]^3\,[PO_4^{3-}] \,/\, [Ag_3PO_4]$
    (e) $K_{sp} \; = \; [Ag^+]\,[PO_4^{3-}]^3 \,/\, [Ag_3PO_4]$

23. What is the K_{sp} for lead(II) fluoride, PbF_2, if the lead ion concentration in a saturated solution is 0.0019 M?
    (a) $K_{sp} \; = \; 1.4 \times 10^{-5}$
    (b) $K_{sp} \; = \; 7.2 \times 10^{-6}$
    (c) $K_{sp} \; = \; 1.4 \times 10^{-8}$
    (d) $K_{sp} \; = \; 2.7 \times 10^{-8}$
    (e) $K_{sp} \; = \; 6.9 \times 10^{-9}$

Section 16.9 Equilibria Shifts for Slightly Soluble Compounds

24. Which of the changes listed below will shift the equilibrium to the right for the following reversible reaction?

$$CaCO_3(s) \quad \leftrightarrow \quad Ca^{2+}(aq) \; + \; CO_3^{2-}(aq)$$

    (a) increase $[Ca^{2+}]$
    (b) increase $[CO_3^{2-}]$
    (c) add solid $Ca(NO_3)_2$
    (d) add solid $NaNO_3$
    (e) decrease pH

25. Which of the changes listed below has no effect on the equilibrium for the following reversible reaction?

$$Hg_2I_2(s) \quad \leftrightarrow \quad Hg_2^{2+}(aq) \; + \; 2\,I^-(aq)$$

    (a) increase $[Hg_2^{2+}]$
    (b) increase $[I^-]$
    (c) add solid Hg_2I_2
    (d) add solid $Hg_2(NO_3)_2$
    (e) add solid KI

Chapter 16 Key Terms

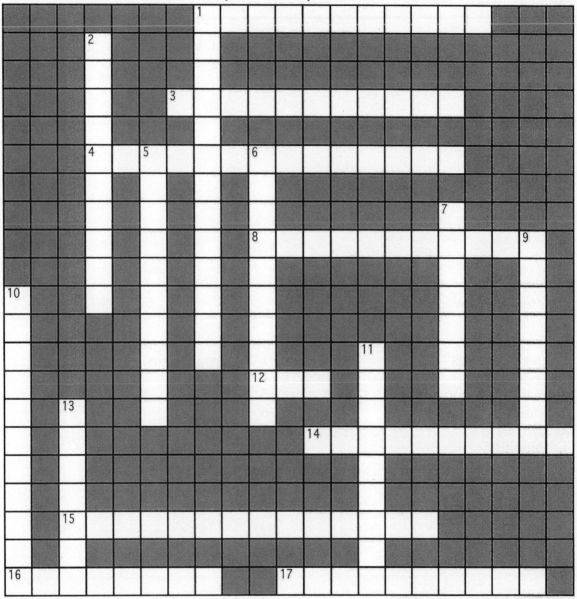

Across

1. an equilibrium with species in the same physical state
3. stated the effect of stress on a chemical equilibrium
4. energy of reactants – energy of products (3 words)
8. a chemical reaction that consumes heat energy
12. a general principle
14. a constant relating molarity of ions in solution
15. the speed that concentrations change (3 words)
16. a particle containing two or more nonmetal atoms
17. the state of highest energy on the reaction profile

Down

1. an equilibrium with species in different states
2. a chemical reaction that liberates heat energy
5. the energy required to reach the transition state
6. a reaction that proceeds toward reactants or products
7. a graph of the energy for a reaction
9. a substance that lowers the energy of activation
10. a dynamic state for a reversible reaction
11. a theory to explain the rate of a chemical reaction
13. an equilibrium constant for all reversible reactions

Chapter 16 Answers to Self-Test

1. d   2. e   3. a   4. c   5. e   6. c   7. d   8. e   9. b   10. c   11. b   12. d   13. a   14. e   15. e
16. d   17. e   18. a   19. c   20. d   21. b   22. c   23. d   24. e   25. c

Chapter 16 Answers to Crossword Puzzle

Across
1. homogeneous
3. LeChatelier
4. heat of reaction
8. endothermic
12. law
14. solubility
15. rate of reaction
16. molecule
17. transition

Down
1. heterogeneous
2. exothermic
5. activation
6. reversible
7. profile
9. catalyst
10. equilibrium
11. collision
13. general

Oxidation and Reduction

Section 17.1 *Oxidation Numbers*

1. What is the oxidation number of iron in a length of wire?
   (a)   0
   (b)   +2
   (c)   +3
   (d)   all of the above
   (e)   none of the above

2. What is the oxidation number of greenish-yellow chlorine gas in the free state?
   (a)   0
   (b)   −1
   (c)   +1
   (d)   +3
   (e)   +5

3. What is the oxidation number of mercury in the mercurous ion, Hg_2^{2+} ?
   (a)   0
   (b)   +1
   (c)   +2
   (d)   +4
   (e)   none of the above

4. What is the oxidation number of chromium in the dichromate ion, $Cr_2O_7^{2-}$?
   (a)   −2
   (b)   +2
   (c)   +6
   (d)   +7
   (e)   none of the above

5. What is the oxidation number of carbon in $Pb(HCO_3)_4$?
   (a)   −4
   (b)   +2
   (c)   +3
   (d)   +4
   (e)   none of the above

6. What is the oxidation number of chlorine in aqueous $HClO_3$?
   (a) –1
   (b) +1
   (c) +3
   (d) +5
   (e) none of the above

Section 17.2 *Oxidation–Reduction Reactions*

7. What substance is oxidized in the following redox reaction?

$$Cl_2(g) \quad + \quad 2\,Br^-(aq) \quad \rightarrow \quad 2\,Cl^-(aq) \quad + \quad Br_2(l)$$

   (a) Cl_2
   (b) Br^-
   (c) Cl^-
   (d) Br_2
   (e) H_2O

8. What substance is the oxidizing agent in the preceding redox reaction?
   (a) Cl_2
   (b) Br^-
   (c) Cl^-
   (d) Br_2
   (e) H_2O

Section 17.3 *Balancing Redox Equations: Oxidation Number Method*

9. After balancing the following redox reaction, what is the coefficient of Mn?

$$Fe_2O_3(l) \quad + \quad Mn(l) \quad \rightarrow \quad MnO_2(l) \quad + \quad Fe(l)$$

   (a) 1
   (b) 2
   (c) 3
   (d) 4
   (e) none of the above

10. After balancing the following redox reaction, what is the coefficient of H^+?

$$MnO_4^-(aq) + VO_2^-(aq) + H^+(aq) \quad \rightarrow \quad Mn^{2+}(aq) + VO_3^-(aq) + H_2O(l)$$

   (a) 2
   (b) 3
   (c) 5
   (d) 6
   (e) none of the above

11. Which of the following is *not* a guideline for balancing redox equations by the oxidation number method?
    (a) Diagram the electrons lost by the substance oxidized and gained by the substance reduced.
    (b) Write a half-reaction for the substance oxidized and the substance reduced.
    (c) Place a coefficient in front of the substance oxidized that corresponds to the number of electrons gained by the substance reduced.
    (d) Place a coefficient in front of the substance reduced that corresponds to the number of electrons lost by the substance oxidized.
    (e) Verify the total number of atoms and total ionic charge is equal for reactants and products.

Section 17.4 *Balancing Redox Equations: Half-Reaction Method*

12. Which of the following is *not* a guideline for balancing redox equations in acid by the half-reaction method?
    (a) Write a half-reaction for the substance oxidized and the substance reduced.
    (b) Balance the atoms in each half-reaction; balance oxygen with water and hydrogen with H^+.
    (c) Multiply each half-reaction by a whole number so that the number of electrons lost by the substance oxidized is equal to the electrons gained by the substance reduced.
    (d) Add the two half-reactions together and cancel identical species on each side of the equation.
    (e) Verify that the total number of atoms is equal to the total ionic charge.

13. Which of the following is *not* a guideline for balancing redox equations in base by the half-reaction method?
    (a) Write a half-reaction for the substance oxidized and the substance reduced.
    (b) Balance the atoms in each half-reaction; balance oxygen with water and hydrogen with OH^-.
    (c) Multiply each half-reaction by a whole number so that the number of electrons lost by the substance oxidized is equal to the electrons gained by the substance reduced.
    (d) Add the two half-reactions together and cancel identical species on each side of the equation.
    (e) Verify that the total ionic charge is equal for reactants and products.

14. After balancing the following redox reaction in acidic solution, what is the coefficient of water?

$$MnO_4^-(aq) \; + \; Co^{2+}(aq) \quad \rightarrow \quad Mn^{2+}(aq) \; + \; Co^{3+}(aq)$$

    (a) 2
    (b) 3
    (c) 4
    (d) 5
    (e) none of the above

15. After balancing the following redox reaction in basic solution, what is the coefficient of water?

$$MnO_4^-(aq) \ + \ Co^{2+}(aq) \quad \rightarrow \quad MnO_2(s) \ + \ Co^{3+}(aq)$$

(a) 2
(b) 3
(c) 5
(d) 6
(e) none of the above

Section 17.5 *Predicting Spontaneous Redox Reactions*

16. Which of the substances listed below is the *strongest* oxidizing agent given the following *spontaneous* redox reaction?

$$FeCl_3(aq) \ + \ NaI(aq) \quad \rightarrow \quad I_2(s) \ + \ FeCl_2(aq) \ + \ NaCl(aq)$$

(a) $FeCl_3$
(b) NaI
(c) I_2
(d) $FeCl_2$
(e) NaCl

17. Which of the substances listed below is the *strongest* reducing agent given the following *nonspontaneous* redox reaction?

$$FeCl_3(aq) \ + \ NaBr(aq) \quad \rightarrow \quad Br_2(l) \ + \ FeCl_2(aq) \ + \ NaCl(aq)$$

(a) $FeCl_3$
(b) NaBr
(c) Br_2
(d) $FeCl_2$
(e) NaCl

18. Given that the following redox reactions go essentially to completion, which of the metals listed below has the *greatest* tendency to undergo oxidation?

$$
\begin{array}{lcll}
Co(s) \ + \ Sn^{2+}(aq) & \rightarrow & Sn(s) \ + & Co^{2+}(aq) \\
Al(s) \ + \ Zn^{2+}(aq) & \rightarrow & Zn(s) \ + & Al^{3+}(aq) \\
Sn(s) \ + \ Ag^+(aq) & \rightarrow & Ag(s) \ + & Sn^{2+}(aq) \\
Zn(s) \ + \ Co^{2+}(aq) & \rightarrow & Co(s) \ + & Zn^{2+}(aq)
\end{array}
$$

(a) Ag
(b) Co
(c) Sn
(d) Al
(e) Zn

19. Given that the above redox reactions go essentially to completion, which of the metals listed below is the *strongest* oxidizing agent?
    (a) Ag
    (b) Co
    (c) Sn
    (d) Al
    (e) Zn

Section 17.6 *Voltaic Cells*

20. Which of the statements listed below is true regarding the following redox reaction occurring in a *spontaneous* electrochemical cell?

$$Cr(s) \quad + \quad Ni^{2+}(aq) \quad \rightarrow \quad Ni(s) \ + \ Cr^{3+}(aq)$$

    (a) Cr is oxidized at the anode
    (b) Ni^{2+} is reduced at the anode
    (c) electrons flow from the Ni electrode to the Cr electrode
    (d) anions in the salt bridge flow from the Cr half-cell to the Ni half-cell
    (e) all of the above

21. Which of the statements listed below is true regarding the following redox reaction occurring in a *spontaneous* electrochemical cell?

$$F_2(g) \quad + \quad 2\,Br^-(aq) \quad \rightarrow \quad Br_2(l) \ + \ 2\,F^-(aq)$$

    (a) Br^- is oxidized at the anode
    (b) F_2 is reduced at the cathode
    (c) electrons flow from the anode to the cathode
    (d) anions in the salt bridge flow from the F_2 half-cell to the Br_2 half-cell
    (e) all of the above

22. Nickel–cadmium batteries are used in rechargeable electronic calculators. Given the *spontaneous* reaction for a discharging nicad battery, what substance is being *reduced*?

$$Cd(s) \ + \ NiO_2(s) \ + \ 2\,H_2O(l) \quad \rightarrow \quad Cd(OH)_2(s) \ + \ Ni(OH)_2(s)$$

    (a) Cd
    (b) NiO_2
    (c) H_2O
    (d) $Cd(OH)_2$
    (e) $Ni(OH)_2$

23. Which of the following is an example of a spontaneous electrochemical cell that has a commercial application as a common voltaic cell?
    (a) alkaline dry cell
    (b) mercury dry cell
    (c) lead storage battery
    (d) nicad battery
    (e) all of the above

24. Which of the statements listed below is true regarding the following redox reaction occurring in a *nonspontaneous* electrochemical cell?

$$MgCl_2(l) \xrightarrow{\text{electricity}} Mg(l) + Cl_2(g)$$

   (a) magnesium chloride is produced at the anode
   (b) magnesium metal is produced at the cathode
   (c) reduction half-reaction: $2\,Cl^- \rightarrow Cl_2 + 2\,e^-$
   (d) oxidation half-reaction: $Mg^{2+} + 2\,e^- \rightarrow Mg$
   (e) none of the above

25. Which of the statements listed below is true regarding the following redox reaction in a *nonspontaneous* electrochemical cell?

$$Br_2(l) + 2\,NaCl(aq) \xrightarrow{\text{electricity}} Cl_2(g) + 2\,NaBr(aq)$$

   (a) liquid bromine is reduced at the anode
   (b) gaseous chlorine is oxidized at the cathode
   (c) electrons flow from the anode to the cathode
   (d) bromine is a stronger oxidizing agent than chlorine
   (e) none of the above

Chapter 17 Key Terms

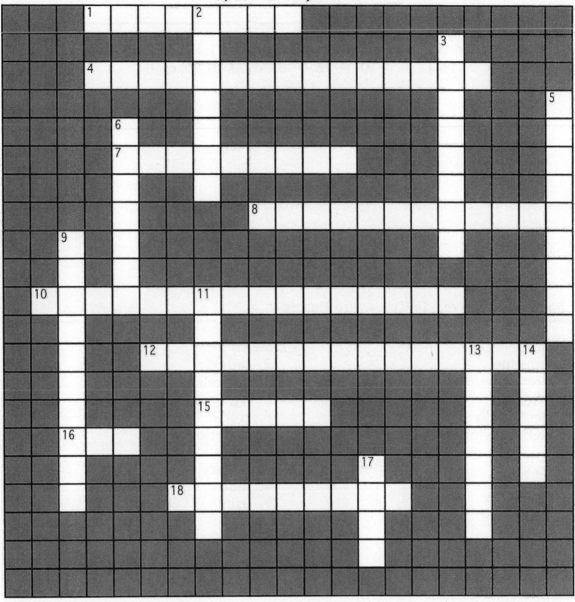

Across

1. the agent undergoing oxidation
4. a cell having two electrodes connected by a wire
7. the agent undergoing reduction
8. a nonspontaneous electrochemical cell
10. the conversion of chemical and electrical energy
12. indicates an electron rich or poor atom (2 words)
15. the electrode at which oxidation occurs
16. an electrochemical cell without aqueous solution
18. the ability of a substance to undergo reduction

Down

2. the electrode at which reduction occurs
3. a cell having a single electrode
5. a process in which a substance gains electrons
6. a spontaneous electrochemical cell
9. allows ions to move between half-cells (2 words)
11. a process in which a substance loses electrons
13. a term for any device that produces electricity
14. a reaction in which electron transfer occurs
17. a reaction involving either oxidation or reduction

Chapter 17 Answers to Self-Test

1. a 2. a 3. b 4. c 5. d 6. d 7. b 8. a 9. c 10. d 11. b 12. e 13. b 14. c 15. a
16. a 17. d 18. d 19. a 20. a 21. e 22. b 23. e 24. b 25. c

Chapter 17 Answers to Crossword Puzzle

Across
1. reducing
4. electrochemical
7. oxidizing
8. electrolytic
10. electrochemistry
12. oxidation number
15. anode
16. dry
18. potential

Down
2. cathode
3. half-cell
5. reduction
6. voltaic
9. salt bridge
11. oxidation
13. battery
14. redox
17. half

Nuclear Chemistry

Section 18.1 *Natural Radioactivity*

1. What are the three main types of radioactive emission from an atomic nucleus?
    (a) proton, neutron, and electron
    (b) proton, neutron, and positron
    (c) positron, neutron, and electron
    (d) positron, neutron, and electron
    (e) alpha, beta, and gamma

2. Which type of nuclear radiation is identical to a helium nucleus and is deflected toward the negative electrode as it passes between electrically charged plates?
    (a) alpha
    (b) beta
    (c) gamma
    (d) all of the above
    (e) none of the above

3. Which type of natural radiation can pass through the human body and requires thick lead protective shielding?
    (a) alpha
    (b) beta
    (c) gamma
    (d) all of the above
    (e) none of the above

Section 18.2 *Nuclear Equations*

4. What is the name of the particle having the following atomic notation: $^{4}_{2}He$?
    (a) alpha
    (b) beta
    (c) gamma
    (d) neutron
    (e) positron

5. What is the name of the radiation having the following atomic notation: $_{0}^{0}\gamma$?
   (a) alpha
   (b) beta
   (c) gamma
   (d) neutron
   (e) positron

6. What is the name of the particle having the following atomic notation: $_{0}^{1}n$?
   (a) alpha
   (b) electron
   (c) neutron
   (d) positron
   (e) proton

7. What is the approximate mass and relative charge of a beta particle?
   (a) 4 amu and 2+
   (b) 1 amu and 1+
   (c) 1 amu and 0
   (d) 0 amu and 1–
   (e) 0 amu and 1+

8. What is the approximate mass and relative charge of a positron?
   (a) 4 amu and 2+
   (b) 1 amu and 1+
   (c) 1 amu and 0
   (d) 0 amu and 1–
   (e) 0 amu and 1+

9. What is the approximate mass and relative charge of a proton?
   (a) 4 amu and 2+
   (b) 1 amu and 1+
   (c) 1 amu and 0
   (d) 0 amu and 1–
   (e) 0 amu and 1+

10. What nuclide is produced when a K-43 nucleus decays by beta emission?

    (a) $_{18}^{43}Ar$ (b) $_{19}^{42}K$ (c) $_{20}^{42}Ca$ (d) $_{20}^{43}Ca$ (e) $_{20}^{44}Ca$

11. What nuclide is produced when an O-15 nucleus decays by positron emission?

    (a) $_{6}^{14}C$ (b) $_{7}^{14}N$ (c) $_{7}^{15}N$ (d) $_{8}^{14}O$ (e) $_{9}^{15}F$

12. What nuclide is produced when a Ba-133 nucleus decays by electron capture?

    (a) $_{55}^{132}Cs$ (b) $_{55}^{133}Cs$ (c) $_{56}^{132}Ba$ (d) $_{57}^{132}La$ (e) $_{57}^{133}La$

Section 18.3 *Radioactive Decay Series*

13. The uranium-238 decay series begins with the emission of an alpha particle. If the daughter decays by beta emission, what is the resulting particle?

    (a) $^{234}_{90}Th$
    (b) $^{233}_{91}Pa$
    (c) $^{234}_{91}Pa$
    (d) $^{234}_{93}Np$
    (e) none of the above

14. In the final step of the uranium-238 disintegration series, the parent nuclide decays into lead-206 and an alpha particle. What is the parent nuclide?

    (a) $^{202}_{80}Hg$
    (b) $^{210}_{83}Bi$
    (c) $^{206}_{84}Po$
    (d) $^{210}_{84}Po$
    (e) none of the above

Section 18.4 *Radioactive Half-Life*

15. Each "click" registered by a Geiger counter indicates which of the following?
    (a) one half-life
    (b) one scintillation
    (c) one minute elapsing
    (d) one second elapsing
    (e) one nucleus decaying

16. The initial activity of a radionuclide was 1000 dpm and dropped to 125 dpm after 72 days. What is the half-life of the radionuclide?
    (a) $t_{1/2} = 18$ days
    (b) $t_{1/2} = 24$ days
    (c) $t_{1/2} = 36$ days
    (d) $t_{1/2} = 144$ days
    (e) $t_{1/2} = 216$ days

17. If 100 mg of a radionuclide is used for medical diagnosis, how much of the radionuclide is still radioactive after 30 hours? *($t_{1/2} = 15$ hours)*
    (a) 25 mg
    (b) 33 mg
    (c) 50 mg
    (d) 100 mg
    (e) 300 mg

18. If the initial activity of a radionuclide is 160 dpm, how much time elapses before the activity drops to 10 dpm? ($t_{1/2} = $ *48 minutes*)
    (a) 12 min
    (b) 96 min
    (c) 144 min
    (d) 192 min
    (e) 768 min

Section 18.5 *Applications of Radionuclides*

19. An archaeologist discovers an artifact with a carbon-14 activity of 7.6 dpm per gram of carbon. If carbon-14 has an activity of 15.3 dpm, what is the estimated age of the parchment? ($t_{1/2} = $ *5730 years*)
    (a) 1400 years
    (b) 2900 years
    (c) 5700 years
    (d) 11,000 years
    (e) 17,000 years

20. What radionuclide can be used to irradiate food and destroy microorganisms?
    (a) cobalt-60
    (b) iodine-131
    (c) iridium-192
    (d) sodium-24
    (e) xenon-133

Section 18.6 *Induced Radioactivity*

21. Bombarding sodium-23 with a proton produces particle X and a neutron. What is particle X ?
    (a) neon-23
    (b) sodium-24
    (c) magnesium-23
    (d) magnesium-24
    (e) none of the above

22. Firing an accelerated particle at a boron-10 target nucleus produces nitrogen-14 and a gamma ray. What is the projectile particle?
    (a) alpha
    (b) beta
    (c) neutron
    (d) proton
    (e) positron

Section 18.7 *Nuclear Fission*

23. A single neutron causes uranium-235 to fission and release three neutrons. Assuming each of the neutrons causes a second fission that releases two neutrons, how many neutrons are released in the second step?
    (a) 3
    (b) 6
    (c) 8
    (d) 9
    (e) none of the above

24. How many neutrons are produced from the following fission reaction?

$$^{235}_{92}U + {}^{1}_{0}n \rightarrow {}^{137}_{52}Te + {}^{96}_{40}Zr + {}^{1}_{0}n$$

    (a) 1
    (b) 2
    (c) 3
    (d) 4
    (e) none of the above

Section 18.8 *Nuclear Fusion*

25. Which of the following best describes the process of nuclear fusion?
    (a) Combining two light nuclei into a heavy nucleus at 25 °C.
    (b) Combining two light nuclei into a heavy nucleus at ~10,000,000 °C.
    (c) Combining two light nuclei into a heavy nucleus at ~1000 K.
    (d) Splitting a heavy nucleus into two lighter nuclei at 25 °C.
    (e) Splitting a heavy nucleus into two lighter nuclei at ~10,000,000 °C.

26. The nuclear fusion of deuterium and tritium gives a neutron and particle-X. Identify particle-X.
    (a) hydrogen-3
    (b) helium-3
    (c) helium-4
    (d) lithium-5
    (e) none of the above

Nuclear Equations for Transmutation Reactions

27. Phosphorus-30 was the first synthetic radionuclide and was produced by bombarding aluminum-27 with an alpha particle. Write a balanced nuclear equation for the reaction.

28. Element 106 was first synthesized in Germany by firing an oxygen-18 projectile at a californium-249 target. Write a balanced nuclear equation for the reaction if the nuclide produced had a mass number of 263.

Nuclear Equations for Fission and Fusion Reactions

29. A neutron can produce a fission reaction of uranium-235 to give xenon-142, strontium-91, and neutrons. Write a balanced nuclear equation for the reaction.

30. Deuterium and tritium undergo a fusion reaction to produce helium-4 and a neutron at 40,000,000 K. Write a balanced nuclear equation for the reaction.

Chapter 18 Key Terms

Across

2. the rate of decay of radionuclides
4. a nucleus of a specific radioactive isotope
6. a reaction in which two light nuclei combine
9. the minimum mass that sustains a chain reaction
11. the nuclide resulting from a decaying nucleus
14. the emission of particles from an unstable nucleus
16. a nuclear radiation identical to an electron
17. a fissionable atom having a mass number of 239
18. the nuclide of hydrogen having one neutron
19. the A value refers to the _____ number
21. a description of a change involving a nucleus
24. an accounting of protons and neutrons for a reaction
25. a positive electron
26. a nuclear radiation identical to high-energy light

Down

1. a nucleus of a specific isotope
3. the nuclide of hydrogen having two neutrons
5. the Z value refers to the _____ number
7. a reaction in which a nucleus splits
8. the life of 50% of the radionuclides in a sample
10. a fissionable atom having a mass number of 235
12. the conversion of one element into another
13. a nuclide that decays to give a daughter product
15. a heavy nuclide draws an electron into its nucleus
20. the stepwise disintegration of a radionuclide
22. a fission reaction that initiates additional fissions
23. a nuclear radiation identical to a helium-4 nucleus

Chapter 18 Answers to Self-Test

1. **e** 2. **a** 3. **c** 4. **a** 5. **c** 6. **c** 7. **d** 8. **e** 9. **b** 10. **d** 11. **c** 12. **b** 13. **c** 14. **d** 15. **e**
16. **b** 17. **a** 18. **d** 19. **c** 20. **a** 21. **c** 22. **a** 23. **b** 24. **c** 25. **b** 26. **c**

Nuclear Equations for Transmutation Reactions

27. $^{27}_{13}Al + ^{4}_{2}He \rightarrow ^{30}_{15}P + ^{1}_{0}n$

28. $^{249}_{98}Cf + ^{18}_{8}O \rightarrow ^{263}_{106}Sg + 4\,^{1}_{0}n$

29. $^{235}_{92}U + ^{1}_{0}n \rightarrow ^{142}_{54}Xe + ^{91}_{38}Sr + 3\,^{1}_{0}n$

30. $^{2}_{1}H + ^{3}_{1}H \rightarrow ^{4}_{2}He + ^{1}_{0}n$

Chapter 18 Answers to Crossword Puzzle

Across
2. activity
4. radionuclide
6. fusion
9. critical
11. daughter
14. radioactivity
16. beta
17. plutonium
18. deuterium
19. mass
21. reaction
24. equation
25. positron
26. gamma

Down
1. nuclide
3. tritium
5. atomic
7. fission
8. half
10. uranium
12. transmutation
13. parent
15. capture
20. series
22. chain
23. alpha

Organic Chemistry

Section 19.1 *Hydrocarbons*

1. Which of the following is a class of saturated hydrocarbons?
    (a) alkanes
    (b) alkenes
    (c) alkynes
    (d) aromatic
    (e) none of the above

Section 19.2 *Alkanes*

2. What is the IUPAC name of the following alkane:

    $$CH_3-CH_2-CH_2-CH_2-CH_2-CH_2-CH_2-CH_2-CH_3?$$

    (a) decane
    (b) heptane
    (c) hexane
    (d) nonane
    (e) none of the above

3. How many isomers share the molecular formula C_6H_{14}?
    (a) 2
    (b) 3
    (c) 4
    (d) 5
    (e) 6

4. What is the name of the following alkyl group: $(CH_3)_2CH-$?
    (a) butyl
    (b) ethyl
    (c) isopropyl
    (d) methyl
    (e) propyl

5. Which of the following is the structural formula for 2-methylhexane?
    (a) $CH_3–CH(CH_3)–CH_2–CH_2–CH_3$
    (b) $CH_3–CH(CH_3)–CH_2–CH_2–CH_2–CH_3$
    (c) $CH_3–CH_2–CH(CH_3)–CH_2–CH_2–CH_3$
    (d) $CH_3–CH(CH_3)–CH_2–CH_2–CH_2–CH_2–CH_3$
    (e) $CH_3–CH_2–CH(CH_3)–CH_2–CH_2–CH_2–CH_3$

6. Which of the following is a balanced equation for the complete combustion of propane?

    (a) $2\,C_3H_8 \;+\; 3\,O_2 \quad \xrightarrow{\text{spark}} \quad 6\,CO \;+\; 8\,H_2$

    (b) $2\,C_3H_8 \;+\; 7\,O_2 \quad \xrightarrow{\text{spark}} \quad 6\,CO \;+\; 8\,H_2O$

    (c) $C_3H_8 \;+\; 3\,O_2 \quad \xrightarrow{\text{spark}} \quad 3\,CO_2 \;+\; 4\,H_2$

    (d) $C_3H_8 \;+\; 5\,O_2 \quad \xrightarrow{\text{spark}} \quad 3\,CO_2 \;+\; 4\,H_2O$

    (e) $C_3H_8 \;+\; O_2 \quad \xrightarrow{\text{spark}} \quad$ no reaction

Section 19.3 *Alkenes and Alkynes*

7. Which of the following is the condensed structural formula for 3-octene?
    (a) $CH_3–CH_2–CH=CH–CH_2–CH_2–CH_3$
    (b) $CH_3–CH_2–CH_2–CH=CH–CH_2–CH_3$
    (c) $CH_3–CH_2–CH_2–CH_2–CH=CH–CH_2–CH_3$
    (d) $CH_3–CH_2–CH_2–CH=CH–CH_2–CH_2–CH_3$
    (e) none of the above

8. What is the IUPAC name of the following: $CH_3–CH_2–C{\equiv}CH$?
    (a) butyne
    (b) 1-butyne
    (c) 2-butyne
    (d) 3-butyne
    (e) none of the above

9. What are the products from the complete combustion of acetylene?
    (a) C_2H_2 and H_2O
    (b) CO and H_2O
    (c) CO and H_2
    (d) CO_2 and H_2O
    (e) CO_2 and H_2

10. Which of the following illustrates the hydrogenation of propylene gas with nickel catalyst at 25 °C and 1 atm?

(a) $CH_2=CH-CH_3 + H_2 \xrightarrow{\text{Ni}} C_3H_4$

(b) $CH_2=CH-CH_3 + H_2 \xrightarrow{\text{Ni}} C_3H_6$

(c) $CH_2=CH-CH_3 + H_2 \xrightarrow{\text{Ni}} C_3H_8$

(d) $CH_2=CH-CH_3 + H_2 \xrightarrow{\text{Ni}} C_6H_6$

(e) $CH_2=CH-CH_3 + H_2 \xrightarrow{\text{Ni}}$ no reaction

Section 19.4 *Arenes*

11. How many representations of benzene are there according to the delocalized electron model?
    (a) 1
    (b) 2
    (c) 3
    (d) 4
    (e) none of the above

12. How many different isomers are there for dimethylbenzene, $C_6H_4(CH_3)_2$?
    (a) 0
    (b) 2
    (c) 3
    (d) 4
    (e) none of the above

Section 19.5 *Hydrocarbon Derivatives*

13. Which of the following classes of compounds does *not* have a carbonyl group?
    (a) aldehyde
    (b) amide
    (c) ester
    (d) ketone
    (e) phenol

14. What is the general formula for an organic halide?
    (a) R–X
    (b) R–OH
    (c) Ar–OH
    (d) R–O–R′
    (e) R–NH$_2$

15. What is the general formula for an alcohol?
    (a) R–X
    (b) R–OH
    (c) Ar–OH
    (d) R–O–R′
    (e) R–NH$_2$

16. What is the general formula for an ether?
    (a) R–X
    (b) R–OH
    (c) Ar–OH
    (d) R–O–R′
    (e) R–NH$_2$

17. What class of compound has the following general formula: $R-\overset{\overset{\displaystyle O}{\|}}{C}-H$?
    (a) aldehyde
    (b) ketone
    (c) carboxylic acid
    (d) ester
    (e) amide

18. What class of compound has the following general formula: $R-\overset{\overset{\displaystyle O}{\|}}{C}-OH$?
    (a) aldehyde
    (b) ketone
    (c) carboxylic acid
    (d) ester
    (e) amide

19. What class of compound is the following: C_6H_5–OH?
    (a) organic halide
    (b) alcohol
    (c) phenol
    (d) ether
    (e) amine

20. What class of compound is the following: CH_3CH_2–NH$_2$?
    (a) organic halide
    (b) alcohol
    (c) phenol
    (d) ether
    (e) amine

Section 19.6 *Organic Halides*

21. What is the condensed structural formula for ethyl bromide?
    (a) $CH_3–Br$
    (b) $CH_3–CH_2–Br$
    (c) $CH_3–CH_2–CH_2–Br$
    (d) $CH_3–CHBr–CH_3$
    (e) none of the above

Section 19.7 *Alcohols, Phenols, and Ethers*

22. What is the condensed structural formula for isopropyl alcohol?
    (a) $CH_3–CH_2–OH$
    (b) $CH_3–CH_2–CH_2–OH$
    (c) $CH_3–CH(OH)–CH_3$
    (d) $CH_3–CH(OH)–CH_2–CH_3$
    (e) none of the above

23. What is the common name for the following: $CH_3CH_2–O–CH_3$?
    (a) diethyl ether
    (b) ethyl ether
    (c) ethyl methyl ether
    (d) methyl ether
    (e) none of the above

Section 19.8 *Amines*

24. What is the condensed structural formula for propylamine?
    (a) $CH_3–CH_2–NH_2$
    (b) $CH_3–CH_2–CH_2–NH_2$
    (c) $CH_3–CH(NH_2)–CH_3$
    (d) $CH_3–CH(NH_2)–CH_2–CH_3$
    (e) none of the above

Section 19.9 *Aldehydes and Ketones*

25. Which of the following organic compounds is most likely an aldehyde based upon the nomenclature suffix?
    (a) chlordane (insecticide)
    (b) cholesterol (steroid)
    (c) citral (lemon fragrance)
    (d) codeine (narcotic)
    (e) cresol (antiseptic)

26. What is the IUPAC systematic name for the following: $CH_3—\overset{\overset{\displaystyle O}{||}}{C}—CH_3$?
    (a)   acetone
    (b)   dimethyl ketone
    (c)   propanaldehyde
    (d)   propanal
    (e)   propanone

Section 19.10 *Carboxylic Acids, Esters, and Amides*

27. What is the IUPAC systematic name for the following: $CH_3—\overset{\overset{\displaystyle O}{||}}{C}—OH$?
    (a)   acetic acid
    (b)   ethanoic acid
    (c)   formic acid
    (d)   methanoic acid
    (e)   propanoic acid

28. What is the IUPAC systematic name for the following: $H—\overset{\overset{\displaystyle O}{||}}{C}—O—CH_2CH_3$?
    (a)   ethyl formate
    (b)   ethyl methanoate
    (c)   ethyl propanoate
    (d)   methyl acetate
    (e)   methyl ethanoate

IUPAC Nomenclature of Organic Compounds

29. What is the IUPAC name for the following compound?

$$CH_3–CH_2–CH–CH_2–CH–CH_2–CH_3$$
$$\qquad\qquad\ |\qquad\qquad\ |$$
$$\qquad\quad CH_3\qquad CH_3$$

30. What is the IUPAC name for the following compound?

$$CH_2{=}CH–CH_2–CH–CH_2–CH_3$$
$$\qquad\qquad\qquad\ |$$
$$\qquad\qquad CH_2–CH_3$$

Chapter 19 Key Terms

Across

1. a family of saturated hydrocarbons
3. a compound containing C, H, and O
6. a hydrocarbon containing all single bonds
10. a carbon and oxygen double bond
11. the second member of the alkane family
12. a compound containing only hydrogen and carbon
17. same molecular formula but different structures
19. a formula showing the molecular arrangement
20. an alkane fragment after removing a H atom
21. a family of compounds
22. the first member of the alkane family

Down

1. a reaction of a hydrocarbon with H_2 or Br_2
2. a group typical of a class of compounds
4. a family of aromatic hydrocarbons
5. a long-chain composed of joined small molecules
7. an arene fragment after removing a H atom
8. the functional group in a carboxylic acid
9. a benzene fragment after removing a H atom
10. a reaction of a hydrocarbon with O_2
12. the functional group in an alcohol or phenol
13. the compounds containing carbon
14. a family of compounds with a double bond
15. a hydrocarbon containing a double or triple bond
16. a hydrocarbon containing a benzene ring
18. a family of compounds with a triple bond

Chapter 19 Answers to Self-Test

1. **a** 2. **d** 3. **d** 4. **c** 5. **b** 6. **d** 7. **c** 8. **b** 9. **d** 10. **c** 11. **a** 12. **c** 13. **e** 14. **a** 15. **b**
16. **d** 17. **a** 18. **c** 19. **c** 20. **e** 21. **b** 22. **c** 23. **c** 24. **b** 25. **c** 26. **e** 27. **b** 28. **b**

IUPAC Nomenclature of Organic Compounds

29. 3,5-dimethylheptane

30. 4-ethyl-1-hexene

Chapter 19 Answers to Crossword Puzzle

Across
1. alkanes
3. derivative
6. saturated
10. carbonyl
11. ethane
12. hydrocarbon
17. isomers
19. structural
20. alkyl
21. class
22. methane

Down
1. addition
2. functional
4. arenes
5. polymer
7. aryl
8. carboxyl
9. phenyl
10. combustion
12. hydroxyl
13. organic
14. alkenes
15. unsaturated
16. aromatic
18. alkynes

Biochemistry

Section 20.1 *Biological Compounds*

1. What type of biological compound is a polymer of amino acids?
    (a) protein
    (b) carbohydrate
    (c) lipid
    (d) nucleic acid
    (e) none of the above

2. What type of biological compound is a polymer of simple sugar molecules?
    (a) protein
    (b) carbohydrate
    (c) lipid
    (d) nucleic acid
    (e) none of the above

3. What type of biological compound is characterized by carboxylic acid, alcohol, and ester functional groups?
    (a) protein
    (b) carbohydrate
    (c) lipid
    (d) nucleic acid
    (e) none of the above

4. What type of biological compound is characterized by aldehyde, alcohol, and amine functional groups?
    (a) protein
    (b) carbohydrate
    (c) lipid
    (d) nucleic acid
    (e) none of the above

5. Which of the following types of bonds is found in a protein?
    (a) peptide linkage
    (b) glycoside linkage
    (c) ester linkage
    (d) phosphate linkage
    (e) none of the above

6.   Which of the following types of bonds is found in a carbohydrate?
     (a)   peptide linkage
     (b)   glycoside linkage
     (c)   ester linkage
     (d)   phosphate linkage
     (e)   none of the above

7.   Which of the following types of bonds is found in a lipid?
     (a)   peptide linkage
     (b)   glycoside linkage
     (c)   ester linkage
     (d)   phosphate linkage
     (e)   none of the above

8.   Which of the following types of bonds is found in a nucleic acid?
     (a)   peptide linkage
     (b)   glycoside linkage
     (c)   ester linkage
     (d)   phosphate linkage
     (e)   none of the above

Section 20.2 *Proteins*

9.   What is the overall shape of a protein that serves a structural function, such as movement or locomotion?
     (a)   compact and spherical
     (b)   long and extended
     (c)   random shape
     (d)   all of the above
     (e)   none of the above

10. What is the overall shape of a protein that serves a metabolic role, such as oxygen transport?
     (a)   compact and spherical
     (b)   long and extended
     (c)   random shape
     (d)   all of the above
     (e)   none of the above

11. What is the overall shape of a protein after it undergoes denaturation?
     (a)   compact and spherical
     (b)   long and extended
     (c)   random shape
     (d)   all of the above
     (e)   none of the above

12. What type of bonds are responsible for the secondary structure of a protein?
    (a) amide bonds
    (b) ester bonds
    (c) hydrogen bonds
    (d) ionic bonds
    (e) none of the above

13. What type of protein structure corresponds to a pleated sheet of amino acids?
    (a) primary
    (b) secondary
    (c) tertiary
    (d) quaternary
    (e) none of the above

14. How many amino acids compose valylalanyltyrosine?
    (a) 1
    (b) 2
    (c) 3
    (d) 4
    (e) 1000s

15. How many amino acid sequences are possible for a tripeptide containing proline (Pro), histidine (His), and leucine (Leu)?
    (a) 1
    (b) 3
    (c) 6
    (d) 9
    (e) thousands

Section 20.3 *Enzymes*

16. An enzyme is an example of which type of biological compound?
    (a) protein
    (b) carbohydrate
    (c) lipid
    (d) nucleic acid
    (e) none of the above

17. Which of the following occurs in Step 1 of enzyme catalysis?
    (a) The substrate cleaves the enzyme molecule.
    (b) The enzyme cleaves the substrate molecule.
    (c) The substrate molecule binds to the active site.
    (d) The enzyme molecule binds to the active site.
    (e) none of the above

18. Which of the following occurs in Step 2 of enzyme catalysis?
    (a) The substrate cleaves the enzyme molecule.
    (b) The enzyme cleaves the substrate molecule.
    (c) The substrate molecule binds to the active site.
    (d) The enzyme molecule binds to the active site.
    (e) none of the above

19. What is the term for a molecule that blocks the active site on an enzyme?
    (a) catalyst
    (b) inhibitor
    (c) steroid
    (d) substrate
    (e) none of the above

Section 20.4 *Carbohydrates*

20. Which of the following functional groups are found in a ketopentose?
    (a) alcohol and aldehyde
    (b) alcohol and ketone
    (c) aldehyde and ketone
    (d) aldehyde and phenol
    (e) none of the above

21. How many carbon atoms are in a molecule of ketopentose sugar?
    (a) 1
    (b) 5
    (c) 6
    (d) 12
    (e) none of the above

22. Which monosaccharides are produced from the acid hydrolysis of sucrose?
    (a) fructose and glucose
    (b) galactose and glucose
    (c) gulose and glucose
    (d) ribose and glucose
    (e) none of the above

23. What is the repeating monosaccharide in cellulose?
    (a) fructose
    (b) galactose
    (c) glucose
    (d) ribose
    (e) none of the above

Section 20.5 *Lipids*

24. Which of the following are characteristic of a lipid fat?
    (a) liquid
    (b) plant source
    (c) unsaturated fatty acids
    (d) water-insoluble
    (e) all of the above

25. Which of the following is *not* found in a fat or oil?
    (a) ester linkages
    (b) glycerol
    (c) saturated fatty acids
    (d) steroid
    (e) unsaturated fatty acids

26. Which of the following is the structure of palmitic acid?
    (a) $CH_3-(CH_2)_{10}-COOH$
    (b) $CH_3-(CH_2)_{12}-COOH$
    (c) $CH_3-(CH_2)_{14}-COOH$
    (d) $CH_3-(CH_2)_{16}-COOH$
    (e) none of the above

27. What are the products from the saponification of trilinolein?
    (a) water and sodium linoleate
    (b) water and linoleic acid
    (c) glycerol and linoleic acid
    (d) glycerol and sodium linoleate
    (e) none of the above

28. What are the products from the saponification of a lipid wax?
    (a) long-chain alcohol and a fatty acid
    (b) long-chain alcohol and a sodium salt of a fatty acid
    (c) glycerol and a fatty acid
    (d) glycerol and a sodium salt of a fatty acid
    (e) none of the above

Section 20.6 *Nucleic Acids*

29. Which of the following is found in both DNA and RNA nucleotides?
    (a) deoxyribose sugar
    (b) phosphoric acid
    (c) ribose sugar
    (d) thymine
    (e) uracil

30. How many nucleotide sequences are possible for a DNA trinucleotide having a cytosine base and two guanine bases?
    (a) 1
    (b) 3
    (c) 6
    (d) 9
    (e) none of the above

31. During DNA replication, an adenine base on the template DNA strand will code for which base in the complementary DNA strand?
    (a) cytosine
    (b) guanine
    (c) thymine
    (d) uracil
    (e) the base varies

32. During RNA transcription, an adenine base on the template DNA strand will code for which base on the growing RNA strand?
    (a) cytosine
    (b) guanine
    (c) thymine
    (d) uracil
    (e) the base varies

Protein Synthesis

33. Explain the process of protein synthesis, and the role of DNA and RNA. Define the term "codon."

Chapter 20 Key Terms

Across

1. the study of biological compounds and reactions
4. a type of bond that joins two amino acids
5. an ester of a long-chain alcohol and a fatty acid
6. symbolizes the products in the enzyme model
7. an acid with a long hydrocarbon side-chain
11. a lipid containing four rings of carbon atoms
12. abbreviation for the amino acid tryptophan
16. abbreviation for the amino acid leucine
18. a carbohydrate polymer of simple sugars
20. a carboxylic acid containing an $-NH_2$ group
22. a water-insoluble compound such as a fat or oil
23. abbreviation for the amino acid phenylalanine
24. abbreviation for one of the acidic amino acids
25. a lipid that contains mostly unsaturated fatty acids
26. a compound that catalyzes a biochemical reaction
28. an acid that carries genetic information
30. an $-O-$ bond that joins two simple sugars

Down

2. a reaction of a triglyceride and sodium hydroxide
3. a polyhydroxy aldehyde or ketone
4. 50 amino acids linked in a long-chain molecule
7. a lipid that contains mostly saturated fatty acids
8. an ester of glycerol and three fatty acids
9. symbolizes the enzyme in the enzyme model
10. a double hydrogen-bonded base pair in DNA
13. a carbohydrate composed of two simple sugars
14. a repeating unit in a nucleic acid
15. an ester of glycerol, 2 fatty acids, and H_3PO_4
17. the process by which DNA synthesizes RNA
18. a polymer of amino acids
19. the process by which DNA synthesizes DNA
21. two amino acids joined by an amide bond
24. abbreviation for the amino acid alanine
27. a triple hydrogen-bonded base pair in DNA
29. abbreviation for one of the basic amino acids

1. **a** 2. **b** 3. **c** 4. **d** 5. **a** 6. **b** 7. **c** 8. **d** 9. **b** 10. **a** 11. **c** 12. **c** 13. **b** 14. **c** 15. **c** 16. **a** 17. **c** 18. **b** 19. **b** 20. **b** 21. **b** 22. **a** 23. **c** 24. **d** 25. **d** 26. **c** 27. **d** 28. **b** 29. **b** 30. **b** 31. **c** 32. **d**

Protein Synthesis

33. First, DNA transcribes genetic code onto a molecule of RNA in the cell nucleus. That is, a single strand of DNA acts as the template for a complementary series of nucleotides and assembles the nucleotides in a molecule of RNA. Next, RNA moves out of the cell nucleus into the cytoplasm where it provides instructions for the sequence of amino acids in a growing protein chain. Three consecutive base nucleotides (called a "codon") specify a given amino acid. For example, the RNA trinucleotide GCG codes for alanine, and AUG codes for methionine.

Chapter 20 Answers to Crossword Puzzle

Across
1. biochemistry
4. peptide
5. wax
6. lock
7. fatty
11. steroid
12. Trp
16. Leu
18. polysaccharide
20. amino
22. lipid
23. Phe
24. Asp
25. oil
26. enzyme
28. nucleic
30. glycoside

Down
2. saponification
3. carbohydrate
4. polypeptide
7. fat
8. triglyceride
9. key
10. AT (A=T)
13. disaccharide
14. nucleotide
15. phospholipid
17. transcription
18. protein
19. replication
21. dipeptide
24. Ala
27. GC (G≡C)
29. Lys

Selected Solutions Manual

Odd-Numbered Exercises

for

Introductory Chemistry
CONCEPTS AND CRITICAL THINKING

Introduction to Chemistry

Section 1.1 *Evolution of Chemistry*

1. The Chinese believed that two forces, *yin* and *yang*, were responsible for bringing the earthly world and everything in nature into existence.

3. Empedocles stated *air, earth, fire,* and *water* were four basic elements responsible for the natural world.

5. Step 1 of the scientific method is to plan and perform an experiment, then make observations and record data.

7. Step 3 of the scientific method is to conduct additional experiments in order to test the hypothesis.

9. A scientific *theory* explains the behavior of nature using a model or concept, whereas a *natural law* states a relationship that is measurable under various experimental conditions.

11. Statement (a) is a *natural law* because mass can be measured.
    Statement (b) is a *theory* because gas molecule collisions cannot be measured.
    Statement (c) is a *natural law* because gas volumes can be measured.
    Statement (d) is a *theory* because it is a model that explains the nucleus.

Section 1.2 *Modern Chemistry*

13. Chemistry is referred to as the central science because it supports other physical sciences, life sciences, and earth sciences. An understanding of chemistry is also prerequisite to the health sciences.

15. Scientists and engineers must have formal training in chemistry, as well as doctors, nurses, dentists, physical therapists, chiropractors, dietitians, and veterinarians.

Section 1.3 *Learning Chemistry*

17. A solution to the nine-dot problem using three straight lines is shown below; the assumption regards the angle of the lines and the size of the dots.

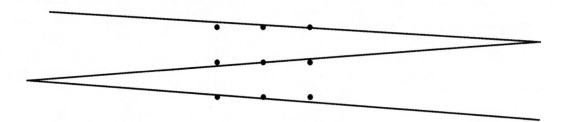

19. By staring at the point where the blocks intersect, we can "flip" the image and view the blocks as stacking upward, or hanging downward.

Scientific Measurements

CHAPTER

2

Section 2.1 *Uncertainty in Measurements*

1.

	Unit	Quantity		Unit	Quantity
(a)	centimeter	length	(b)	gram	mass
(c)	milliliter	volume	(d)	second	time

3. (a) 10.0 cm and (d) 10.5 cm each have an uncertainty of ± 0.1 cm.

5. (c) 75.22 g is the only mass with an uncertainty of ± 0.01 g.

7. (b) 25.0 mL and (c) 25.5 mL each have an uncertainty of ± 0.5 mL.

Section 2.2 *Significant Digits*

9.

	Measurement	Significant Digits
(a)	0.05 cm	1 significant digit
(b)	0.50 cm	2 significant digits
(c)	25.0 cm	3 significant digits
(d)	20.50 cm	4 significant digits

11.

	Measurement	Significant Digits
(a)	0.5 mL	1 significant digit
(b)	0.50 mL	2 significant digits
(c)	5.00 mL	3 significant digits
(d)	500 mL	1 significant digit

13.

	Measurement	Significant Digits
(a)	1.050×10^2 cm	4 significant digits
(b)	2×10^3 cm	1 significant digit
(c)	3.00×10^{-4} cm	3 significant digits
(d)	5.0×10^{-5} cm	2 significant digits

Section 2.3 *Rounding Off Nonsignificant Digits*

15.	Example	Rounded Off
(a)	10.25	10.3
(b)	10.20	10.2
(c)	0.01029	0.0103
(d)	10,248	10,200

17.	Example	Rounded Off
(a)	1.454×10^1	1.45×10^1
(b)	1.455×10^2	1.46×10^2
(c)	1.508×10^{-3}	1.51×10^{-3}
(d)	1.503×10^{-4}	1.50×10^{-4}

Section 2.4 *Adding and Subtracting Measurements*

19. (a) 1.55 cm + 36.15 cm +17.3 cm = 55.00 cm   *rounds to 55.0 cm*
     (b) 5.0 cm + 16.3 cm + 0.95 cm = 22.25 cm   *rounds to 22.3 cm*

21. (a) 242.167 g – 175 g = 67.167 g   *rounds to 67 g*
     (b) 27.55 g – 14.545 g = 13.005 g   *rounds to 13.01 g*

Section 2.5 *Multiplying and Dividing Measurements*

23. (a) 3.65 cm × 2.10 cm = 7.665 cm^2 *rounds to 7.67 cm^2*
     (b) 8.75 cm × 1.15 cm = 10.0625 cm^2 *rounds to 10.1 cm^2*
     (c) 16.5 cm × 1.7 cm = 28.05 cm^2 *rounds to 28 cm^2*
     (d) 21.1 cm × 20 cm = 422 cm^2 *rounds to 400 cm^2*

25. (a) $\dfrac{26.0 \text{ cm}^2}{10.1 \text{ cm}}$ = 2.5742 cm   *rounds to 2.57 cm*

     (b) $\dfrac{9.95 \text{ cm}^3}{0.15 \text{ cm}^2}$ = 66.333 cm   *rounds to 66 cm*

     (c) $\dfrac{131.78 \text{ cm}^3}{19.25 \text{ cm}}$ = 6.8457 cm^2 *rounds to 6.846 cm^2*

     (d) $\dfrac{131.78 \text{ cm}^3}{19.2 \text{ cm}}$ = 6.8635 cm^2 *rounds to 6.86 cm^2*

Section 2.6 *Exponential Numbers*

27. (a) $10 \times 10 \times 10 \times 10 = 10^4$

     (b) $\dfrac{1}{10} \times \dfrac{1}{10} \times \dfrac{1}{10} = (\dfrac{1}{10})^3 = 10^{-3}$

29.   (a) $2 \times 2 \times 2 = 2^3$

       (b) $\frac{1}{2} \times \frac{1}{2} \times \frac{1}{2} = \left(\frac{1}{2}\right)^3 = 2^{-3}$

31.

	Scientific Notation	Ordinary Number
(a)	1×10^3	1000
(b)	1×10^{-7}	0.000 000 1

33.

	Ordinary Number	Scientific Notation
(a)	1,000,000,000	1×10^9
(b)	0.000 000 01	1×10^{-8}

35.

	Scientific Notation	Ordinary Number
(a)	1×10^1	10
(b)	1×10^{-1}	0.1

Section 2.7 *Scientific Notation*

37.

	Ordinary Number	Scientific Notation
(a)	1,010,000,000,000,000	1.01×10^{15}
(b)	0.000 000 000 000 456	4.56×10^{-13}
(c)	94,500,000,000,000,000	9.45×10^{16}
(d)	0.000 000 000 000 000 019 50	1.950×10^{-17}

39. 2.69×10^{22} helium atoms

41. 3.35×10^{-23} g/neon atom

Section 2.8 *Unit Equations and Unit Factors*

43.

	Unit Equation
(a)	1 mile = 5280 feet
(b)	1 ton = 2000 pounds

45.   Unit Factors

    (a) $\dfrac{1 \text{ mile}}{5280 \text{ feet}}$ and $\dfrac{5280 \text{ feet}}{1 \text{ mile}}$

    (b) $\dfrac{1 \text{ ton}}{2000 \text{ pounds}}$ and $\dfrac{2000 \text{ pounds}}{1 \text{ ton}}$

47. (a) 1 mile = 5280 feet is an exact equivalent;
    (b) 1 mile = 1.61 kilometers is an approximate equivalent because it relates two different systems of measurement (English and metric).

49. (a) 1 liter = 1000 milliliters is an exact equivalent;
    (b) 1 quart = 946 milliliters is an approximate equivalent because it relates two different systems of measurement (English and metric).

51. (a) 1 mile = 5280 feet has an infinite number of significant digits because it is an exact equivalent;
    (b) 1 mile = 1.61 kilometers has three significant digits because it is an approximate equivalent.

53. (a) 1 liter = 1000 milliliters has an infinite number of significant digits because it is an exact equivalent;
    (b) 1 quart = 946 milliliters has three significant digits because it is an approximate equivalent.

Section 2.9 *Unit Analysis Problem Solving*

55. $36.0 \ \text{inches} \ \times \ \dfrac{2.54 \ \text{centimeters}}{1 \ \text{inches}} \ = \ 91.4 \ \text{centimeters}$

57. $0.750 \ \text{carat} \ \times \ \dfrac{0.200 \ \text{gram}}{1 \ \text{carat}} \ = \ 0.150 \ \text{grams}$

59. $2.5 \ \text{pints} \ \times \ \dfrac{2 \ \text{cups}}{1 \ \text{pint}} \ = \ 5.0 \ \text{cups}$

61. $5.5 \ \text{years} \ \times \ \dfrac{12 \ \text{months}}{1 \ \text{year}} \ = \ 66 \ \text{months}$

63. $275{,}000{,}000{,}000 \ \text{troy ounces} \ \times \ \dfrac{1 \ \text{troy pound}}{12 \ \text{troy ounces}} \ = \ 2.29 \times 10^{10} \ \text{troy pounds}$

65. $1.00 \ \text{hour} \ \times \ \dfrac{60 \ \text{minutes}}{1 \ \text{hour}} \ \times \ \dfrac{60 \ \text{seconds}}{1 \ \text{minute}} \ \times \ \dfrac{3.00 \times 10^{8} \ \text{m}}{1 \ \text{second}} \ = \ 1.08 \times 10^{12} \ \text{m}$

Section 2.10 *The Percent Concept*

67. $65.5 \ \text{g sterling silver} \ \times \ \dfrac{7.50 \ \text{g copper}}{100 \ \text{g sterling silver}} \ = \ 4.91 \ \text{g copper}$

69.   255 mL ethanol + 375 mL water = 630 mL = 6.30×10^2 mL solution

Note: Since 630 mL represents two significant digits, we must express the total as 6.30×10^2 mL.

$$\frac{255 \text{ mL}}{6.30 \times 10^2 \text{ mL}} \times 100\% = 40.5\% \text{ of solution is ethanol}$$

71. $15.0 \text{ g oxygen} \times \dfrac{100 \text{ g water}}{88.8 \text{ g oxygen}} = 16.9 \text{ g water}$

73.   Oxygen: $2.37 \times 10^{25} \text{ g crust} \times \dfrac{49.2 \text{ g oxygen}}{100 \text{ g crust}} = 1.17 \times 10^{25}$ g oxygen

Silicon: $2.37 \times 10^{25} \text{ g crust} \times \dfrac{25.7 \text{ g silicon}}{100 \text{ g crust}} = 6.09 \times 10^{24}$ g silicon

Aluminum: $2.37 \times 10^{25} \text{ g crust} \times \dfrac{7.50 \text{ g aluminum}}{100 \text{ g crust}} = 1.78 \times 10^{24}$ g aluminum

75.   Mass of coin = 3.051 g (copper and zinc)
Mass of zinc = 0.153 g
Mass of copper = 3.051 g – 0.153 g = 2.898 g

Percent of copper: $\dfrac{2.898 \text{ g}}{3.051 \text{ g}} \times 100\% = 94.99\%$ copper

General Exercises

77.   10.00 mL ($\pm$ 0.01 mL)

79.   186,000 miles per second (1.86×10^5 miles per second)

81.   126.457 g + 131.6 g = 258.057 g   *rounds to* 258.1 g

83.

Exponential Number	Scientific Notation
(a) 352×10^4	3.52×10^6
(b) 0.191×10^{-5}	1.91×10^{-6}

85.   Total mass: $(1.6749 \times 10^{-24} \text{ g}) + (1.6726 \times 10^{-24} \text{ g}) = 3.3475 \times 10^{-24}$ g

87.   1 metric ton = 2200 lb = 2.200×10^3 lb
1 English ton = 2000 lb = 2.000×10^3 lb
Mass difference: 2200 lb – 2000 lb = 200 lb
Mass difference: 2.200×10^3 lb – 2.000×10^3 lb = 0.200×10^3 lb = 2.00×10^2 lb

89. 4.0 billion years $= 4.0 \times 10^9$ years

$$4.0 \times 10^9 \; \cancel{\text{years}} \; \times \; \frac{365 \; \cancel{\text{days}}}{1 \; \cancel{\text{year}}} \times \frac{24 \; \text{hours}}{1 \; \cancel{\text{day}}} = 3.5 \times 10^{13} \; \text{hours}$$

91. A+: $\dfrac{18,825 \; \text{patients}}{55,368 \; \text{patients}} \times 100\% = 34.0\%$ A+ blood type

Note: This is not a measurement; thus, significant digits do not apply.

93. $1 \; \cancel{\text{pound}} \; \times \; \dfrac{16 \; \cancel{\text{ounces}}}{1 \; \cancel{\text{pound}}} \; \times \; \dfrac{28.4 \; \text{grams}}{1 \; \cancel{\text{ounce}}} \; = \; 454 \; \text{grams (feathers)}$

95. 1 pound $=$ 454 grams (feathers)
    1 troy pound $=$ 373 grams (gold)

Since 454 grams is greater than 373 grams, *a pound of feathers* weighs more than a pound of gold.

The Metric System

Section 3.1 *Basic Units and Symbols*

1. The (b) gram and (c) liter are basic units; (a) kilometer and (d) millisecond are prefix units in the metric system.

3.

	Unit	Symbol		Unit	Symbol
(a)	kilometer	km	(b)	gigagram	Gg
(c)	deciliter	dL	(d)	picosecond	ps

5.

	Unit	Quantity		Unit	Quantity
(a)	kilometer	length	(b)	gigagram	mass
(c)	deciliter	volume	(d)	picosecond	time

7.

	Instrument	Quantity		Instrument	Quantity
(a)	ruler	length	(b)	platform balance	mass
(c)	pipet	volume	(d)	stopwatch	time

9.

	Prefix	Name/Symbol		Prefix	Name/Symbol
(a)	1×10^{12}	tera (T)	(b)	1×10^9	giga (G)
(c)	1×10^{-3}	milli (m)	(d)	1×10^{-6}	micro (μ)

11.

	Symbol	Unit		Symbol	Unit
(a)	Tm	terameter	(b)	Gg	gigagram
(c)	mL	milliliter	(d)	μs	microsecond

Section 3.2 *Metric Conversion Factors*

13.   <u>Unit Equation</u>   <u>Unit Equation</u>

       (a) 1×10^{12} m = 1 Tm;   (b) 1×10^{9} g = 1 Gg;

       (c)   1 L = 1×10^{3} mL;   (d)   1 s = 1×10^{6} μs

15.   <u>Unit Factors</u>   <u>Unit Factors</u>

       (a) $\dfrac{1 \times 10^{12} \text{ m}}{1 \text{ Tm}}$ and $\dfrac{1 \text{ Tm}}{1 \times 10^{12} \text{ m}}$ (b) $\dfrac{1 \times 10^{9} \text{ g}}{1 \text{ Gg}}$ and $\dfrac{1 \text{ Gg}}{1 \times 10^{9}}$

       (c) $\dfrac{1 \text{ L}}{1 \times 10^{3} \text{ mL}}$ and $\dfrac{1 \times 10^{3} \text{ mL}}{1 \text{ L}}$ (d) $\dfrac{1 \text{ s}}{1 \times 10^{6} \text{ } \mu s}$ and $\dfrac{1 \times 10^{6} \text{ } \mu s}{1 \text{ s}}$

Section 3.3 *Metric–Metric Conversions*

17.   (a) $1.00 \; \cancel{\text{m}} \times \dfrac{1 \text{ km}}{1000 \; \cancel{\text{m}}} = 1.00 \times 10^{-3}$ km

      (b) $100 \; \cancel{\text{g}} \times \dfrac{100 \text{ cg}}{1 \; \cancel{\text{g}}} = 1 \times 10^{4}$ cg

      (c) $0.100 \; \cancel{\text{L}} \times \dfrac{10 \text{ dL}}{1 \; \cancel{\text{L}}} = 1.00$ dL

      (d) $0.000\,010 \; \cancel{\text{s}} \times \dfrac{1 \times 10^{9} \text{ ns}}{1 \; \cancel{\text{s}}} = 1.00 \times 10^{4}$ ns

19.   (a) $6.50 \; \cancel{\text{Tm}} \times \dfrac{1 \times 10^{12} \; \cancel{\text{m}}}{1 \; \cancel{\text{Tm}}} \times \dfrac{1 \text{ Mm}}{1 \times 10^{6} \; \cancel{\text{m}}} = 6.50 \times 10^{6}$ Mm

      (b) $650 \; \cancel{\text{Gg}} \times \dfrac{1 \times 10^{9} \; \cancel{\text{g}}}{1 \; \cancel{\text{Gg}}} \times \dfrac{1 \text{ kg}}{1000 \; \cancel{\text{g}}} = 6.50 \times 10^{8}$ kg

      (c) $0.650 \; \cancel{\text{cL}} \times \dfrac{1 \; \cancel{\text{L}}}{100 \; \cancel{\text{cL}}} \times \dfrac{10 \text{ dL}}{1 \; \cancel{\text{L}}} = 6.50 \times 10^{-2}$ dL

      (d) $0.\,000\,650 \; \cancel{\text{ns}} \times \dfrac{1 \; \cancel{\text{s}}}{1 \times 10^{9} \; \cancel{\text{ns}}} \times \dfrac{1 \times 10^{12} \text{ ps}}{1 \; \cancel{\text{s}}} = 0.650$ ps

Section 3.4 *Metric–English Conversions*

21.   (a)   2.54 cm = 1 in.   (b)   454 g = 1 lb
       (c)   946 mL = 1 qt   (d)   1.00 s = 1 sec

23.   (a) $86 \; \cancel{\text{cm}} \times \dfrac{1 \text{ in.}}{2.54 \; \cancel{\text{cm}}} = 34 \text{ in.}$

       (b) $36 \; \cancel{\text{g}} \times \dfrac{1 \text{ lb}}{454 \; \cancel{\text{g}}} = 0.079 \text{ lb}$

       (c) $0.500 \; \cancel{\text{qt}} \times \dfrac{946 \text{ mL}}{1 \; \cancel{\text{qt}}} = 473 \text{ mL}$

       (d) $8.00 \times 10^2 \; \cancel{\text{sec}} \times \dfrac{1 \text{ s}}{1.00 \; \cancel{\text{sec}}} = 8.00 \times 10^2 \text{ s}$

25.   (a) $72 \; \cancel{\text{in.}} \times \dfrac{2.54 \; \cancel{\text{cm}}}{1 \; \cancel{\text{in.}}} \times \dfrac{1 \text{ m}}{100 \; \cancel{\text{cm}}} = 1.8 \text{ m}$

       (b) $175 \; \cancel{\text{lb}} \times \dfrac{454 \; \cancel{\text{g}}}{1 \; \cancel{\text{lb}}} \times \dfrac{1 \text{ kg}}{1000 \; \cancel{\text{g}}} = 79.5 \text{ kg}$

       (c) $0.500 \; \cancel{\text{qt}} \times \dfrac{946 \; \cancel{\text{mL}}}{1 \; \cancel{\text{qt}}} \times \dfrac{1 \text{ L}}{1000 \; \cancel{\text{mL}}} = 0.473 \text{ L}$

       (d) $1.05 \times 10^{-4} \; \cancel{\text{min}} \times \dfrac{60 \; \cancel{\text{s}}}{1 \; \cancel{\text{min}}} \times \dfrac{1 \text{ ks}}{1000 \; \cancel{\text{s}}} = 6.30 \times 10^{-6} \text{ ks}$

27. $\dfrac{13 \; \cancel{\text{km}}}{1 \; \cancel{\text{L}}} \times \dfrac{1 \text{ mi}}{1.61 \; \cancel{\text{km}}} \times \dfrac{3.784 \; \cancel{\text{L}}}{1 \text{ gal}} = 31 \text{ mi/gal}$

Section 3.5 *Volume by Calculation*

29. $5.00 \; \cancel{\text{cm}} \times \dfrac{1 \; \cancel{\text{m}}}{100 \; \cancel{\text{cm}}} \times \dfrac{1000 \text{ mm}}{1 \; \cancel{\text{m}}} = 50.0 \text{ mm}$

      $50.0 \text{ mm} \times 50.0 \text{ mm} \times 25.0 \text{ mm} = 62{,}500 \text{ mm}^3 \; (6.25 \times 10^4 \text{ mm}^3)$

31. $\dfrac{(15.3 \text{ cm}^3)}{(4.95 \text{ cm})(2.45 \text{ cm})} = \dfrac{(15.3 \text{ cm} \times \cancel{\text{cm}} \times \cancel{\text{cm}})}{(4.95 \; \cancel{\text{cm}})(2.45 \; \cancel{\text{cm}})} = 1.26 \text{ cm}$

33. (a) $1\ \cancel{L} \times \dfrac{1000\ \text{mL}}{1\ \cancel{L}} = 1000\ \text{mL}$

(b) $1\ \cancel{L} \times \dfrac{1000\ \cancel{mL}}{1\ \cancel{L}} \times \dfrac{1\ \text{cm}^3}{1\ \cancel{mL}} = 1000\ \text{cm}^3$

35. $6.10\ \cancel{L} \times \dfrac{1000\ \cancel{mL}}{1\ \cancel{L}} \times \dfrac{1\ \cancel{cm^3}}{1\ \cancel{mL}} \times \left(\dfrac{1\ \text{in.}}{2.54\ \cancel{cm}}\right)^3 = 372\ \text{in.}^3$

Section 3.6 *Volume by Displacement*

37. volume of piece of jade: $7.5\ \text{mL} - 4.5\ \text{mL} = 3.0\ \text{mL}$

39. volume of hydrogen gas: $255\ \text{mL} - 125\ \text{mL} = 130\ \text{mL}\ (1.30 \times 10^2\ \text{mL})$

Section 3.7 *The Density Concept*

41.

	Substance	Sink or Float		Substance	Sink or Float
(a)	ebony	sink	(b)	bamboo	float

43.

	Gas	Rise or Drop		Gas	Rise or Drop
(a)	helium	rise	(b)	laughing gas	drop

45. (a) $25.0\ \cancel{g} \times \dfrac{1\ \text{mL}}{0.714\ \cancel{g}} = 35.0\ \text{mL}$

(b) $15.0\ \cancel{g} \times \dfrac{1\ \text{mL}}{0.792\ \cancel{g}} = 18.9\ \text{mL}$

47. (a) $7.50\ \cancel{cm^3} \times \dfrac{2.18\ \text{g}}{1\ \cancel{cm^3}} = 16.4\ \text{g}$

(b) $455\ \cancel{cm^3} \times \dfrac{1.715\ \text{g}}{1\ \cancel{cm^3}} = 780\ \text{g}\ (7.80 \times 10^2\ \text{g})$

49. (a) $\dfrac{19.7\ \text{g}}{25.0\ \text{mL}} = 0.788\ \text{g/mL}$

(b) $\dfrac{11.6\ \text{g}}{4.1\ \text{mL}} = 2.8\ \text{g/mL}$

Section 3.8 *Temperature*

51.
Temperature Scale	Freezing Point of Water
(a) Fahrenheit	32 °F
(b) Kelvin	273 K
(c) Celsius	0 °C

53. (a) $(100 - 32)\,\cancel{°F} \times \dfrac{100\ °C}{180\ \cancel{°F}} = 38\ °C$

    (b) $(-215 - 32)\,\cancel{°F} \times \dfrac{100\ °C}{180\ \cancel{°F}} = -137\ °C$

55. (a) $495\ °C + 273 = 768\ K$

    (b) $-185\ °C + 273 = 88\ K$

Section 3.9 *Heat and Specific Heat*

57. Temperature is a measure of the *average* energy in a container and heat is a measure of the *total* energy in a container. (For example, a large chemistry lecture room at 20 °C has more heat energy than a small lecture room at 20 °C, even though the temperature is the same.)

59. $250\ \cancel{g} \times \dfrac{1.00\ cal}{1\ \cancel{g} \times \cancel{°C}} \times (100 - 23)\,\cancel{°C} = 19{,}000\ cal\ \ (1.9 \times 10^4\ cal)$

61. $25.0\ \cancel{g} \times \dfrac{0.108\ cal}{1\ \cancel{g} \times \cancel{°C}} \times (50.0 - 25.0)\,\cancel{°C} = 67.5\ cal$

63. $\dfrac{35.7\ cal}{75.0\ g \times (43.9 - 28.9)\ °C} = 0.0317\ cal/(g \times °C)$

General Exercises

65. $1.5\ \cancel{TB} \times \dfrac{1 \times 10^{12}\ \cancel{Bytes}}{1\ \cancel{TB}} \times \dfrac{1\ MB}{1 \times 10^6\ \cancel{Bytes}} = 1.5 \times 10^6\ MB$

67. (a) infinite number of digits (*exact by definition*) (b) 3 significant digits
    (c) infinite number of digits (*exact by definition*) (d) 3 significant digits

69. $35 \, \text{ms} \times \dfrac{1 \, \text{s}}{1000 \, \text{ms}} \times \dfrac{1 \times 10^6 \, \mu\text{s}}{1 \, \text{s}} = 3.5 \times 10^4 \, \mu\text{s}$

71. $\dfrac{1.86 \times 10^5 \, \text{mi}}{1 \, \text{s}} \times \dfrac{60 \, \text{s}}{1 \, \text{min}} \times \dfrac{60 \, \text{min}}{1 \, \text{h}} \times \dfrac{24 \, \text{h}}{1 \, \text{d}} \times \dfrac{365 \, \text{d}}{1 \, \text{yr}} = 5.87 \times 10^{12} \, \text{mi/yr}$

73. $1500 \, \text{m} \times \dfrac{1 \, \text{km}}{1000 \, \text{m}} \times \dfrac{1 \, \text{mile}}{1.61 \, \text{km}} = 0.93 \, \text{mile}$

Note: The 1500-meter race is shorter than a mile, thus the time is faster.

75. $2.50 \, \text{kg} \times \dfrac{1000 \, \text{g}}{1 \, \text{kg}} \times \dfrac{1000 \, \text{mg}}{1 \, \text{g}} \times \dfrac{1 \, \text{tablet}}{325 \, \text{mg}} = 7690 \, \text{tablets}$

77. $94.0 \, \text{feet} \times \dfrac{1 \, \text{yard}}{3 \, \text{feet}} \times \dfrac{0.914 \, \text{m}}{1 \, \text{yard}} = 28.6 \, \text{m}$

$50.0 \, \text{feet} \times \dfrac{1 \, \text{yard}}{3 \, \text{feet}} \times \dfrac{0.914 \, \text{m}}{1 \, \text{yard}} = 15.2 \, \text{m}$

Playing area of a basketball court: $28.6 \, \text{m} \times 15.2 \, \text{m} = 435 \, \text{m}^2$

79. Volume of metal sample: $57.5 \, \text{mL} - 50.0 \, \text{mL} = 7.5 \, \text{mL}$

Density of metal sample: $\dfrac{37.51 \, \text{g}}{7.5 \, \text{mL}} = 5.0 \, \text{g/mL}$

81. Specific gravity of gasohol: $\dfrac{0.801 \, \text{g/mL}}{1.00 \, \text{g/mL}} = 0.801$

83. $\dfrac{1.00 \, \text{g}}{1 \, \text{mL}} \times \dfrac{1 \, \text{lb}}{454 \, \text{g}} \times \dfrac{1 \, \text{mL}}{1 \, \text{cm}^3} \times \left(\dfrac{2.54 \, \text{cm}}{1 \, \text{in.}}\right)^3 \times \left(\dfrac{12 \, \text{in.}}{1 \, \text{ft}}\right)^3 = 62.4 \, \text{lb/ft}^3$

85. $(-422 - 32) \, °\text{F} \times \dfrac{100 \, °\text{C}}{180 \, °\text{F}} = -252 \, °\text{C}$

$-252 \, °\text{C} + 273 = 21 \, \text{K}$

87. **Heat gain** by water = **Heat loss** by poker

Heat gain by water: $3500 \; \cancel{g} \times \dfrac{1.00 \; \text{cal}}{1 \; \cancel{g} \times \cancel{°C}} \times (51 - 22) \; \cancel{°C} = 1.0 \times 10^5 \; \text{cal}$

Heat loss by poker: $1.0 \times 10^5 \; \cancel{\text{cal}} \times \dfrac{1 \; \text{kcal}}{1000 \; \cancel{\text{cal}}} = 1.0 \times 10^2 \; \text{kcal}$

89. $\dfrac{75.6 \; \cancel{\text{cal}}}{(31.4 - 20.7) \cancel{°C}} \times \dfrac{1 \; \text{g} \times \cancel{°C}}{0.125 \; \cancel{\text{cal}}} = 56.5 \; \text{g}$

91. $\dfrac{35.2 \; \cancel{\text{cal}}}{10.5 \; \cancel{g}} \times \dfrac{1 \; \cancel{g} \times °C}{0.0566 \; \cancel{\text{cal}}} = 59.2 \; °C$

93. $10.0 \; \cancel{g} \times \dfrac{0.0920 \; \text{cal}}{1 \; \cancel{g} \times °C} \times (T_{\text{final}} - 22.7 \; °C) = 27.8 \; \text{cal}$

$T_{\text{final}} - 22.7 \; °C = 27.8 \; \cancel{\text{cal}} \; / \; 0.920 \; \cancel{\text{cal}} / \; °C$

$T_{\text{final}} - 22.7 \; °C = 30.2 \; °C$

$T_{\text{final}} = 30.2 \; °C + 22.7 \; °C = 52.9 \; °C$

95. $\dfrac{13.6 \; \cancel{g}}{1 \; \cancel{\text{mL}}} \times \dfrac{1 \; \text{kg}}{1000 \; \cancel{g}} \times \dfrac{1 \; \cancel{\text{mL}}}{1 \; \cancel{\text{cm}^3}} \times \left(\dfrac{100 \; \cancel{\text{cm}}}{1 \; \text{m}}\right)^3 = 1.36 \times 10^4 \; \text{kg}/\text{m}^3$

97. $V = \pi \, r^2 h$

$r = 1.95 \; \text{cm}$

Note: By definition, the mass of the International Prototype Kilogram is exactly one kilogram (1 kg).

$V = 1 \; \cancel{\text{kg}} \times \dfrac{1000 \; \cancel{g}}{1 \; \cancel{\text{kg}}} \times \dfrac{\text{cm}^3}{21.50 \; \cancel{g}} = 46.51 \; \text{cm}^3$

Given: $\pi r^2 h = 46.51 \text{ cm}^3$

$$h = \frac{46.51 \text{ cm}^3}{\pi \cdot r^2} = \frac{46.51 \text{ cm}^3}{3.14 \times (1.95 \text{ cm})^2} = 3.90 \text{ cm}$$

Note: diameter $= 2 \times$ radius $= 2 \times 1.95 \text{ cm} = 3.90 \text{ cm}$

Thus, the diameter and height of the International Prototype Kilogram are equal.

Matter and Energy

CHAPTER

4

Section 4.1 *Physical States of Matter*

1. The *liquid state* has a variable shape and fixed volume.

3. The *liquid state* has a variable shape and compresses negligibly.

5. The *liquid state* has particles closely packed and free to move about.

7.
	Change of State	Term
(a)	gas to liquid	condensing
(b)	solid to liquid	melting

9.
	Change of State	Heat/Cool
(a)	gas to liquid	cooling
(b)	solid to liquid	heating

11. The process of sublimation (solid to gas) requires heating, and the solid absorbs heat in the process.

Section 4.2 *Elements, Compounds, and Mixtures*

13.
	Example	Classification
(a)	sodium metal	element
(b)	chlorine gas	element
(c)	sodium chloride	compound
(d)	salt water	mixture

15.
	Example	Classification
(a)	copper wire	element
(b)	copper ore	heterogeneous mixture
(c)	copper oxide	compound
(d)	brass alloy	homogeneous mixture

Section 4.3 *Names and Symbols of the Elements*

17.

	Element	Symbol			Element	Symbol
(a)	lithium	Li		(b)	argon	Ar
(c)	magnesium	Mg		(d)	manganese	Mn
(e)	beryllium	Be		(f)	silicon	Si
(g)	mercury	Hg		(h)	titanium	Ti

19.

	Symbol	Element			Symbol	Element
(a)	Cl	chlorine		(b)	Ne	neon
(c)	Cd	cadmium		(d)	Ge	germanium
(e)	Co	cobalt		(f)	Ra	radium
(g)	Cr	chromium		(h)	Te	tellurium

21.

	Element	Atomic number			Element	Atomic number
(a)	hydrogen	1		(b)	boron	5
(c)	aluminum	13		(d)	titanium	22
(e)	arsenic	33		(f)	strontium	38
(g)	tin	50		(h)	bismuth	83

Section 4.4 *Metals, Nonmetals, and Semimetals*

23.

	Property	Classification
(a)	dense solid	metal
(b)	low boiling point	nonmetal
(c)	ductile	metal
(d)	reacts with metals	nonmetal

25.

	Property	Classification
(a)	shiny solid	metal
(b)	brittle solid	nonmetal
(c)	low density	nonmetal
(d)	forms alloys	metal

27.

	Element	Classification
(a)	Na	metal
(b)	B	semimetal
(c)	Hg	metal
(d)	Kr	nonmetal

29.

	Element	Classification
(a)	beryllium	metal
(b)	germanium	semimetal
(c)	selenium	nonmetal
(d)	xenon	nonmetal

31.

	Element	Physical State
(a)	Na	solid
(b)	N	gas
(c)	Si	solid
(d)	Kr	gas

33.

	Element	Physical State
(a)	hydrogen	gas
(b)	titanium	solid
(c)	fluorine	gas
(d)	bromine	liquid

Section 4.5 *Compounds and Chemical Formulas*

35.

	Chemical Formula	Chemical Composition
(a)	$C_{12}H_{18}Cl_2N_8OS$	12 carbon atoms 18 hydrogen atoms 2 chlorine atoms 8 nitrogen atoms 1 oxygen atom 1 sulfur atom
(b)	$C_{17}H_{20}N_4O_6$	17 carbon atoms 20 hydrogen atoms 4 nitrogen atoms 6 oxygen atoms

37.

	Chemical Composition	Chemical Formula
(a)	20 carbon, 30 hydrogen, 1 oxygen	vitamin A: $C_{20}H_{30}O$
(b)	31 carbon, 46 hydrogen, 2 oxygen	vitamin K: $C_{31}H_{46}O_2$

39.

	Chemical Formula	Total Atoms
(a)	$CH_3(CH_2)_{16}COOH$	stearic acid: 56 atoms
(b)	$C_3H_5(C_{17}H_{33}COO)_3$	triolein: 167 atoms

41. The mass ratio of the elements in water is constant: 1:8.

Section 4.6 *Physical and Chemical Properties*

43.

	Property		Property
(a)	physical	(b)	physical
(c)	physical	(d)	chemical

45.

	Property			Property
(a)	physical		(b)	chemical
(c)	chemical		(d)	physical

47.

	Property			Property
(a)	chemical		(b)	physical
(c)	physical		(d)	physical

49.

	Property			Property
(a)	chemical		(b)	physical
(c)	physical		(d)	chemical

Section 4.7 *Physical and Chemical Changes*

51.

	Change			Change
(a)	chemical		(b)	physical
(c)	physical		(d)	chemical

53.

	Change			Change
(a)	physical		(b)	chemical
(c)	physical		(d)	chemical

55.

	Change			Change
(a)	physical		(b)	chemical
(c)	physical		(d)	chemical

57.

	Change			Change
(a)	chemical		(b)	physical
(c)	chemical		(d)	physical

Section 4.8 *Conservation of Mass*

59. 5.00 g iron + 2.77 g phosphorus = 7.77 g iron phosphide

61. 0.750 g mercury oxide − 0.695 g mercury = 0.055 g oxygen

Section 4.9 *Potential and Kinetic Energy*

63. Particles in the *gaseous state* have more kinetic energy than the liquid state, which has more kinetic energy than the solid state.

65. If the temperature increases, the kinetic energy will *increase*.

67. If a steel cylinder of helium gas is heated from 50 °C to 100 °C, there is more kinetic energy and the motion of helium atoms will *increase*.

Section 4.10 *Conservation of Energy*

69.   From the law of conservation of energy, 1500 kilocalories of heat energy is required to vaporize the water to steam at 100 °C.

71.   697 calories + 110 calories = 807 calories released

73.   From the law of conservation of energy, 48.0 kilocalories of heat is required to decompose 15.0 g of water into hydrogen and oxygen gases.

75.

Energy Conversion	Forms of Energy
(a) burning coal converts water to steam	chemical → heat
(b) steam drives a turbine	heat → mechanical
(c) a turbine turns a generator	mechanical → mechanical
(d) a generator makes electricity	mechanical → electrical

77.

Energy Conversion	Forms of Energy
(a) hot gases moving pistons	heat → mechanical
(b) pistons powering the crankshaft	mechanical → mechanical
(c) crankshaft turning the generator	mechanical → mechanical
(d) generator charging the battery	mechanical → electrical

General Exercises

79.   A pre-1982 penny is an alloy, and thus an example of a *homogeneous mixture*.

81.   Gasoline is a mixture of compounds, but it is a *homogeneous mixture* because the properties are constant throughout a sample.

83.   The element *germanium* is below silicon in Group IV/14; thus germanium has similar chemical properties and can be substituted for silicon.

85.   The three most abundant elements in the Earth's crust are *oxygen, silicon,* and *aluminum.*

87.

Element	Atomic number
(a) argentum	47
(b) plumbum	82
(c) stannum	50
(d) aurum	79

89.   A *physical change* involves a change of physical state, but not a change of chemical formula.

91.   From the conservation of energy law, *3250 calories* of heat is required to decompose 5.00 g of ammonia into hydrogen gas and nitrogen gas.

93. According to the conservation of mass and energy law, the products will weigh *slightly more* than the reactants. In theory, a small amount of heat energy is converted into mass; in practice, the mass gain is undetectable.

95. On a playground swing the potential energy is greatest when the swing is at its highest point. As the swing travels down, it loses potential energy and gains kinetic energy. Thus, the *potential energy is minimum,* and the *kinetic energy is maximum,* when the swing is at its lowest point.

97. E = energy, m = mass, c = speed of light

Models of the Atom

Section 5.1 *Dalton Model of the Atom*

1. (c) Atoms are indivisible later proved to be invalid with Thomson's discovery of the electron and proton.

3. Dalton relied on the work of (1) *Robert Boyle,* who first proposed the particle nature of gases; (2) *Antoine Lavoisier,* who established the law of conservation of mass; and (3) *Joseph Proust,* who established the law of definite composition.

Section 5.2 *Thomson Model of the Atom*

5. The electron (e^-) is the simplest negative particle.

7. The relative charge on an electron (e^-) is minus one (–1).

9. Raisins in plum pudding are analogous to electrons in atoms.

Section 5.3 *Rutherford Model of the Atom*

11. The approximate size of a nucleus is 1×10^{-13} cm.

13. The relative charge of an electron and proton is –1 and +1, respectively.

15. Electrons are found outside the nucleus.

Section 5.4 *Atomic Notation*

17.

	Isotope	Neutrons		Isotope	Neutrons
(a)	$^{23}_{11}\text{Na}$	$23 - 11 = 12\ n^0$	(b)	$^{27}_{13}\text{Al}$	$27 - 13 = 14\ n^0$
(c)	$^{65}_{30}\text{Zn}$	$65 - 30 = 35\ n^0$	(d)	$^{107}_{47}\text{Ag}$	$107 - 47 = 60\ n^0$

19.

	Isotope	Neutrons		Isotope	Neutrons
(a)	hydrogen-2	$2 - 1 = 1\ n^0$	(b)	carbon-14	$14 - 6 = 8\ n^0$
(c)	cobalt-60	$60 - 27 = 33\ n^0$	(d)	iodine-131	$131 - 53 = 78\ n^0$

21.

Atomic Notation	Atomic Number	Mass Number	Number of Protons	Number of Neutrons	Number of Electrons
$^{11}_{5}\text{B}$	5	11	5	6	5
$^{15}_{7}\text{N}$	7	15	7	8	7
$^{40}_{20}\text{Ca}$	20	40	20	20	20
$^{200}_{80}\text{Hg}$	80	200	80	120	80

23.

(a) $\left(\begin{array}{c} 4\ n^0 \\ 3\ p^+ \end{array}\right)$ 3 e–   (b) $\left(\begin{array}{c} 7\ n^0 \\ 6\ p^+ \end{array}\right)$ 6 e–

(c) $\left(\begin{array}{c} 8\ n^0 \\ 8\ p^+ \end{array}\right)$ 8 e–   (d) $\left(\begin{array}{c} 10\ n^0 \\ 10\ p^+ \end{array}\right)$ 10 e–

Section 5.5 *Atomic Mass*

25. The atomic mass is the weighted average of the individual isotopic masses that occur naturally for an element.

27. The current atomic mass scale reference isotope is carbon-12.

29. ^{23}Na has only one naturally occurring isotope with a mass of 22.99 amu.

31. ^{19}F has only one naturally occurring isotope with a mass of 19.00 amu.

33.
Li-6:	6.015 amu × 0.0742	=	0.446 amu
Li-7:	7.016 amu × 0.9258	=	6.495 amu
	Atomic Mass	=	6.941 amu

35.
Fe-54:	53.940 amu × 0.0582	=	3.14 amu
Fe-56:	55.935 amu × 0.9166	=	51.27 amu
Fe-57:	56.935 amu × 0.0219	=	1.25 amu
Fe-58:	57.933 amu × 0.0033	=	0.19 amu
	Atomic Mass	=	55.85 amu

37. If the average mass of bromine is approximately 80 amu, and one isotope is Br-79, the other isotope must be Br-81.

Section 5.6 *The Wave Nature of Light*

39. Violet light has a shorter wavelength than blue light.

41. Violet light has a higher frequency than blue light.

43. Violet light has higher energy than blue light.

45. A wavelength of 550 nm has a higher frequency than 650 nm.

Section 5.7 *The Quantum Concept*

47. A *photon* represents the quantum or particle nature of light energy.

49.
Example	Spectrum
(a) rainbow	continuous
(b) line spectrum	quantized

51.
Example	Spectrum
(a) metric ruler	continuous
(b) digital laser	quantized

Section 5.8 *Bohr Model of the Atom*

53.   <u>The Bohr Model of the Atom</u>

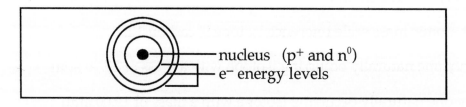

55.   The transition from energy level 5 to 2 is the most energetic because the electron drops from a higher energy level than 4 to 2, or 3 to 2.

57.   <u>Energy Level Change</u>   <u>Color of Emission</u>
            4 to 2   blue-green line

59.   <u>Energy Level Change</u>   <u>Type of Emission</u>
            5 to 1   ultraviolet energy

61.   The *red line* in the emission spectrum of hydrogen has the least energy because a red photon corresponds to the electron dropping the least distance (3 to 2).

63.   <u>Energy Level Change</u>   <u>Number of Photons</u>
       (a)   3 to 1   1
       (b)   3 to 2   1

65.   <u>Energy Level Change</u>   <u>Color</u>
       (a)   2 to 1   ultraviolet
       (b)   3 to 2   red
       (c)   4 to 3   infrared

Section 5.9 *Energy Levels and Sublevels*

67.   The lines in the emission spectrum of hydrogen suggest the existence of energy levels.

69.   <u>Energy Level</u>   <u>Number of Sublevels</u>
       (a)   first level   1 sublevel
       (b)   second level   2 sublevels
       (c)   third level   3 sublevels
       (d)   fourth level   4 sublevels

71.

	Sublevel	Max. Electrons
(a)	2s	2 e–
(b)	4p	6 e–
(c)	3d	10 e–
(d)	5f	14 e–

73. The maximum number of electrons in the second energy level is equal to the sum of the maximum number of electrons in the 2s and 2p sublevels, that is, $2 \text{ e}^- + 6 \text{ e}^- = 8 \text{ e}^-$.

Section 5.10 *Electron Configuration*

75. The 3d sublevel follows the 4s sublevel.

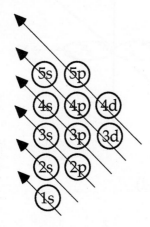

77.

	Element	Electron Configuration
(a)	He	$1s^2$
(b)	Be	$1s^2\ 2s^2$
(c)	Co	$1s^2\ 2s^2\ 2p^6\ 3s^2\ 3p^6\ 4s^2\ 3d^7$
(d)	Cd	$1s^2\ 2s^2\ 2p^6\ 3s^2\ 3p^6\ 4s^2\ 3d^{10}\ 4p^6\ 5s^2\ 4d^{10}$

79.

	Electron Configuration	Element
(a)	$1s^2\ 2s^1$	Li
(b)	$1s^2\ 2s^2\ 2p^6\ 3s^2\ 3p^2$	Si
(c)	$1s^2\ 2s^2\ 2p^6\ 3s^2\ 3p^6\ 4s^2\ 3d^2$	Ti
(d)	$1s^2\ 2s^2\ 2p^6\ 3s^2\ 3p^6\ 4s^2\ 3d^{10}\ 4p^6\ 5s^2$	Sr

81.   An orbit is the path traveled by an electron of given energy about the nucleus of an atom, according to the Bohr model.

An orbital is a region about the nucleus in which there is a high probability of finding an electron of given energy, according to the quantum mechanical model.

83.   (a)   (b)

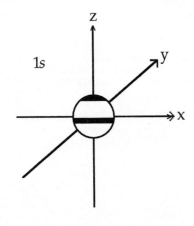

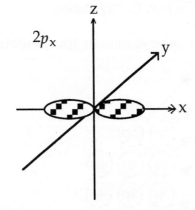

(c)   (d)

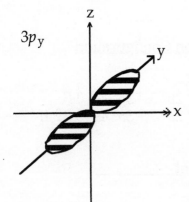

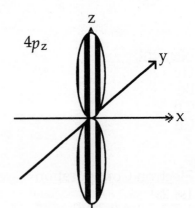

85.

	Orbitals	Higher Energy			Orbitals	Higher Energy
(a)	$2s$ or $3s$	$3s$		(b)	$2p_x$ or $3p_x$	$3p_x$
(c)	$2p_x$ or $2p_y$	both are equal		(d)	$4p_y$ or $4p_z$	both are equal

87.

	Description	Orbital
(a)	spherical orbital in the fifth energy level	$5s$
(b)	dumbbell-shaped orbital in the fourth energy level	$4p$

89.

	Orbital	Max. # of Electrons
(a)	$1s$ orbital	$2\ e^-$
(b)	$2p$ orbital	$2\ e^-$
(c)	$3d$ orbital	$2\ e^-$
(d)	$4f$ orbital	$2\ e^-$

General Exercises

91. $\quad 1.60 \times 10^{-19}\ \text{coulomb} \times \dfrac{1\ g}{1.76 \times 10^8\ \text{coulomb}} = 9.09 \times 10^{-28}\ g$

93. Let X = the mass of Ag-107 isotope
Abundance of Ag-107: $100\% - 48.16\%$ $\quad = \quad 51.84\%$

Ag-107 isotope:	$X \times 0.5184$	$=$	$0.5184\ X$
Ag-109 isotope:	$108.91\ \text{amu} \times 0.4816$	$=$	$52.45\ \text{amu}$
Both isotopes:	*atomic mass*	$=$	$107.87\ \text{amu}$

$$0.5184\ X\ +\ 52.45\ \text{amu} = 107.87\ \text{amu}$$

$$0.5184\ X = 107.87\ \text{amu} - 52.45\ \text{amu}$$

$$0.5184\ X = 55.42\ \text{amu}$$

$$X = \frac{55.42\ \text{amu}}{0.5184}$$

$$\text{mass of Ag-107 isotope} = 106.9\ \text{amu}$$

95. The periodic table lists a mass number for Pm (147), and not an atomic mass. We therefore conclude that there is no stable isotope for promethium.

97.

Wavelength		Region
(a)	200 nm	ultraviolet
(b)	500 nm	visible
(c)	1200 nm	infrared

99. An atom is more stable with a completely filled d sublevel than a partially filled sublevel. If one of the copper $4s$ electrons moves into the $3d$ sublevel, the $3d$ sublevel is completely filled, and thus more stable.

The Periodic Table

Section 6.1 *Classification of Elements*

1. The noble gases were not discovered until 1894–98 (see Figure 4.7).

3. In Figure 6.1, Group IV corresponds to the oxide formula GeO_2, and germanium oxide, GeO_2, is an example.

Section 6.2 *The Periodic Law Concept*

5. After Moseley's discovery in 1913, the periodic law states that physical and chemical properties tend to repeat periodically when elements are arranged according to **increasing atomic number**.

7. Note that **Co** precedes **Ni** in the periodic table, even though the atomic mass of Co (58.93) is greater than Ni (58.69).

Section 6.3 *Groups and Periods of Elements*

9. Horizontal rows in the periodic table are referred to as **periods** or **series**.

11. The term for the elements that belong to Groups IA through VIIIA is the **representative elements**.

13. The elements in the series that follow element 57 are called the **lanthanides**.

15. The term for the two series of elements that belong to Ce–Lu and Th–Lr is the **inner transition series**.

17. The elements on the left side of the periodic table are described as **metals**.

19. The appearance of **B, Si, Ge, As, Sb**, and **Te** resembles that of metals, and these nonradioactive semimetals are referred to as metalloids.

21.

	Family	Group Number
(a)	alkali metals	IA/1
(b)	alkaline earth metals	IIA/2
(c)	halogens	VIIA/17
(d)	noble gases	VIIIA/18

23.

	American	IUPAC		American	IUPAC
(a)	Group IA	1	(b)	Group IIIA	13
(c)	Group VA	15	(d)	Group VIIA	17
(e)	Group IB	11	(f)	Group IIIB	3
(g)	Group VB	5	(h)	Group VIIB	7

25.

	Element		Element
(a)	Ge	(b)	Na
(c)	I	(d)	Pm

27.

	Element		Element
(a)	Mg	(b)	Br
(c)	Lu	(d)	Th

Section 6.4 *Periodic Trends*

29. Proceeding from left to right in the periodic table, the atomic radius of an element generally *decreases*.

31. Proceeding up a group of elements in the periodic table, the atomic radius of an element generally *decreases*.

33.

	Atomic Radius		Atomic Radius
(a)	Li < **Na**	(b)	N < **P**
(c)	Mg < **Ca**	(d)	Ar < **Kr**

(Note: The element with the larger atomic radius is in **bold**.)

35.

	Metallic Character		Metallic Character
(a)	B < **Al**	(b)	Na < **K**
(c)	Mg < **Ba**	(d)	H < **Fe**

(Note: The element with the greater metallic character is in **bold**.)

Section 6.5 Properties of Elements

37. Predicted atomic radius of K: $\quad$ 0.248 nm $-$ (0.266 $-$ 0.248) nm $\;=$ 0.230 nm

Predicted density of Rb: $\quad\dfrac{0.86 \text{ g/mL} + 1.90 \text{ g/mL}}{2} = 1.38 \text{ g/mL}$

Predicted melting point of Cs: $\quad$ 38.9 °C $-$ (63.3 °C $-$ 38.9 °C) $\;=$ 14.5 °C

39. Predicted atomic radius of Cl: $\quad$ 0.115 nm $-$ (0.133 nm $-$ 0.115 nm) $=$ 0.097 nm

Predicted density of Br: $\quad\dfrac{1.56 \text{ g/mL} + 4.97 \text{ g/mL}}{2} = 3.27 \text{ g/mL}$

Predicted boiling point of I: $\quad$ 58.8 °C $+$ [58.8 °C $-$ ($-$34.6 °C)] $\;=$ 152.2 °C

41.

	Compound	Formula		Compound	Formula
(a)	potassium fluoride	KF	(b)	calcium fluoride	CaF_2
(c)	boron bromide	BBr_3	(d)	germanium iodide	GeI_4

43.

	Compound	Formula		Compound	Formula
(a)	calcium fluoride	CaF_2	(b)	calcium chloride	$CaCl_2$
(c)	calcium bromide	$CaBr_2$	(d)	calcium iodide	CaI_2

45.

	Compound	Formula		Compound	Formula
(a)	arsenic oxide	As_2O_3	(b)	antimony oxide	Sb_2O_3
	and,	As_2O_5		and,	Sb_2O_5

Section 6.6 Blocks of Elements

47. The elements in Groups IA/1 and IIA/2 are filling s sublevels.

49. The transition elements are filling d sublevels.

51. The elements in the lanthanide series are filling a $4f$ sublevel.

53.

	Element	Highest Sublevel		Element	Highest Sublevel
(a)	H	$1s$	(b)	Na	$3s$
(c)	Sm	$4f$	(d)	Br	$4p$
(e)	Sr	$5s$	(f)	C	$2p$
(g)	Sn	$5p$	(h)	Cs	$6s$

55.

	Element	Electron Configuration	Core Notation
(a)	Li	$1s^2\,2s^1$	[He] $2s^1$
(b)	F	$1s^2\,2s^2\,2p^5$	[He] $2s^2\,2p^5$
(c)	Mg	$1s^2\,2s^2\,2p^6\,3s^2$	[Ne] $3s^2$
(d)	P	$1s^2\,2s^2\,2p^6\,3s^2\,3p^3$	[Ne] $3s^2\,3p^3$
(e)	Ca	$1s^2\,2s^2\,2p^6\,3s^2\,3p^6\,4s^2$	[Ar] $4s^2$
(f)	Mn	$1s^2\,2s^2\,2p^6\,3s^2\,3p^6\,4s^2\,3d^5$	[Ar] $4s^2\,3d^5$
(g)	Ga	$1s^2\,2s^2\,2p^6\,3s^2\,3p^6\,4s^2\,3d^{10}\,4p^1$	[Ar] $4s^2\,3d^{10}\,4p^1$
(h)	Rb	$1s^2\,2s^2\,2p^6\,3s^2\,3p^6\,4s^2\,3d^{10}\,4p^6\,5s^1$	[Kr] $5s^1$

Section 6.7 *Valence Electrons*

57.

	Group	Valence Electrons		Group	Valence Electrons
(a)	IA/1	1	(b)	IIIA/13	3
(c)	VA/15	5	(d)	VIIA/17	7

59.

	Element	Valence Electrons		Element	Valence Electrons
(a)	H	1	(b)	B	3
(c)	N	5	(d)	F	7
(e)	Ca	2	(f)	Si	4
(g)	O	6	(h)	Ar	8

Section 6.8 *Electron Dot Formulas*

61.

	Element	Electron Dot Formula		Element	Electron Dot Formula
(a)	H	H·	(b)	B	·Ḅ·
(c)	N	·N̈:	(d)	F	:Ḟ:
(e)	Ca	Ça·	(f)	Si	·S̈i·
(g)	O	·Ö:	(h)	Ar	:Är:

Section 6.9 *Ionization Energy*

63. Proceeding from left to right in the periodic table, the ionization energy *increases*.

65. The Group VIIIA/18 elements have the *highest* ionization energy.

67.

	Ionization Energy			Ionization Energy
(a)	**Mg** > Ca		(b)	**S** > Se
(c)	**Sn** > Pb		(d)	**N** > P

(Note: The element with the higher ionization energy is in **bold**.)

69.

	Ionization Energy			Ionization Energy
(a)	Rb > **Cs**		(b)	He > **Ar**
(c)	B > **Al**		(d)	F > **I**

(Note: The element with the lower ionization energy is in **bold**.)

Section 6.10 *Ionic Charges*

71.

	Group	Ionic Charge		Group	Ionic Charge
(a)	IA/1	1+	(b)	IIA/2	2+
(c)	IIIA/13	3+	(d)	IVA/14	4+

73.

	Ion	Ionic Charge		Ion	Ionic Charge
(a)	Cs ion	1+	(b)	Ga ion	3+
(c)	O ion	2–	(d)	I ion	1–

75.

	Ion	Isoelectric		Ion	Isoelectric
(a)	Al^{3+}	Ne	(b)	Ca^{2+}	Ar
(c)	S^{2-}	Ar	(d)	N^{3-}	Ne

77.

	Ion	Electron Configuration – Core Notation
(a)	Mg^{2+}	[Ne]
(b)	K^+	[Ar]
(c)	Fe^{2+}	[Ar] $3d^6$
(d)	Zr^{2+}	[Kr] $4d^2$

79.

	Ion	Electron Configuration – Core Notation
(a)	F^-	[He] $2s^2\ 2p^6$, or [Ne]
(b)	S^{2-}	[Ne] $3s^2\ 3p^6$, or [Ar]
(c)	N^{3-}	[He] $2s^2\ 2p^6$, or [Ne]
(d)	I^-	[Kr] $5s^2\ 4d^{10}\ 5p^6$, or [Xe]

General Exercises

81.   The element Mendeleev predicted and called ekaboron, we now call scandium (Sc), which is named for Scandinavia, its place of discovery.

83.

		European	American			European	American
(a)		Group IA	Group IA	(b)		Group IB	Group IB
(c)		Group IIIA	Group IIIB	(d)		Group IIIB	Group IIIA

85.   Predicted atomic radius of Fr: $0.266 \text{ nm} + (0.266 - 0.248) \text{ nm} = 0.284 \text{ nm}$

Predicted density of Fr: $1.87 \text{ g/mL} + (1.87 - 1.53) \text{ g/mL} = 2.21 \text{ g/mL}$

Predicted melting point of Fr: $28.4 \,°C - (38.9 - 28.4) \,°C = 17.9 \,°C$

87.

	Element	Electron Configuration
(a)	Sr	$[\text{Kr}] \, 5s^2$
(b)	Ru	$[\text{Kr}] \, 5s^2 \, 4d^6$
(c)	Sb	$[\text{Kr}] \, 5s^2 \, 4d^{10} \, 5p^3$
(d)	Cs	$[\text{Xe}] \, 6s^1$

89.   When an alkali metal atom loses one electron, it assumes a noble gas electron configuration. When an alkaline earth metal atom loses one electron, it does not assume a noble gas electron configuration. However, when an alkaline earth metal atom loses two electrons, it does attain a noble gas electron configuration and becomes very stable.

91.   Although hydrogen and the Group IA/1 metals each have one valence electron, in hydrogen the electron is in the $1s$ sublevel and in the Group IA/1 metals the electron is in a higher s sublevel. As a result, in hydrogen the electron is closer to its nucleus than the valence electron in the Group IA/1 metals. Because the negatively charged electron is closer to the positively charged nucleus, it requires more energy to remove the electron from a hydrogen atom than in the other Group IA/1 metals.

Language of Chemistry

CHAPTER

7

Section 7.1 *Classification of Compounds*

1.

	Compound	Classification		Compound	Classification
(a)	NH_3	binary molecular	(b)	$LiNO_3$	ternary ionic
(c)	FeN	binary ionic	(d)	$HNO_3(aq)$	ternary oxyacid

3.

	Ion	Classification		Ion	Classification
(a)	NH_4^+	polyatomic cation	(b)	Ni^{2+}	monoatomic cation
(c)	Br^-	monoatomic anion	(d)	CO_3^{2-}	polyatomic anion

Section 7.2 *Monoatomic Ions*

5.

	Monoatomic Cation	Stock System Name
(a)	K^+	potassium ion
(b)	Ba^{2+}	barium ion
(c)	Fe^{2+}	iron(II) ion
(d)	Co^{3+}	cobalt(III) ion

7.

	Monoatomic Cation	Chemical Formula
(a)	lithium ion	Li^+
(b)	aluminum ion	Al^{3+}
(c)	lead(II) ion	Pb^{2+}
(d)	nickel(II) ion	Ni^{2+}

9.

	Monoatomic Cation	Latin System Name
(a)	Cu^+	cuprous ion
(b)	Fe^{3+}	ferric ion
(c)	Sn^{4+}	stannic ion
(d)	Pb^{2+}	plumbous ion

11.   Monoatomic Cation   Chemical Formula
      (a)   cobaltous ion Co^{2+}
      (b)   cobaltic ion Co^{3+}

13.   Monoatomic Anion   Systematic Name
      (a) F^- fluoride ion
      (b) Cl^- chloride ion
      (c) Br^- bromide ion
      (d) I^- iodide ion

Section 7.3 *Polyatomic Ions*

15.   Oxyanion   Systematic Name
      (a) ClO_3^- chlorate ion
      (b) NO_2^- nitrite ion
      (c) CO_3^{2-} carbonate ion
      (d) HSO_4^{2-} hydrogen sulfate ion

17.   Polyatomic Ion   Chemical Formula
      (a)   dichromate ion $Cr_2O_7^{2-}$
      (b)   phosphate ion PO_4^{3-}
      (c)   hydroxide ion OH^-
      (d)   hypochlorite ion ClO^-

Section 7.4 *Writing Chemical Formulas*

19.   Constituent Ions   Chemical Formula
      (a) Li^+ + Cl^- $LiCl$
      (b) $2 Ag^+$ + O^{2-} Ag_2O
      (c) Cr^{3+} + $3 I^-$ CrI_3
      (d) $3 Sn^{2+}$ + $2 N^{3-}$ Sn_3N_2

21.   Constituent Ions   Chemical Formula
      (a) K^+ + NO_3^- KNO_3
      (b) $2 NH_4^+$ + $Cr_2O_7^{2-}$ $(NH_4)_2Cr_2O_7$
      (c) $2 Al^{3+}$ + $3 SO_3^{2-}$ $Al_2(SO_3)_3$
      (d) Cr^{3+} + $3 ClO^-$ $Cr(ClO)_3$

23.   Constituent Ions   Chemical Formula
      (a) Sr^{2+} + $2 NO_2^-$ $Sr(NO_2)_2$
      (b) Zn^{2+} + $2 MnO_4^-$ $Zn(MnO_4)_2$
      (c) Ca^{2+} + CrO_4^{2-} $CaCrO_4$
      (d) Cr^{3+} + $3 ClO_4^-$ $Cr(ClO_4)_3$

Section 7.5 *Binary Ionic Compounds*

25.   <u>Binary Ionic Compound</u>   <u>Systematic Name</u>
- (a) MgO magnesium oxide
- (b) ZnO zinc oxide
- (c) CdO cadmium oxide
- (d) BaO barium oxide

27.   <u>Binary Ionic Compound</u>   <u>Systematic Name</u>
- (a) $LiBr$ lithium bromide
- (b) SrI_2 strontium iodide
- (c) Na_3N sodium nitride
- (d) AlF_3 aluminum fluoride

29.   <u>Binary Ionic Compound</u>   <u>Chemical Formula</u>
- (a)   copper(I) oxide Cu_2O
- (b)   iron(III) nitride FeN
- (c)   mercuric(II) chloride $HgCl_2$
- (d)   lead(IV) sulfide PbS_2

31.   <u>Binary Ionic Compound</u>   <u>Chemical Formula</u>
- (a)   rubidium chloride $RbCl$
- (b)   sodium bromide $NaBr$

33.   <u>Binary Ionic Compound</u>   <u>Chemical Formula</u>
- (a)   gallium nitride GaN
- (b)   aluminum arsenide $AlAs$

Section 7.6 *Ternary Ionic Compounds*

35.   <u>Ternary Ionic Compound</u>   <u>Systematic Name</u>
- (a) $LiMnO_4$ lithium permanganate
- (b) $Sr(ClO_2)_2$ strontium chlorite
- (c) $BaCO_3$ barium carbonate
- (d) $(NH_4)_2Cr_2O_7$ ammonium dichromate

37.   <u>Ternary Ionic Compound</u>   <u>Stock System Name</u>
- (a) $CuSO_4$ copper(II) sulfate
- (b) $FeCrO_4$ iron(II) chromate
- (c) $Hg(NO_2)_2$ mercury(II) nitrite
- (d) $Pb(C_2H_3O_2)_2$ lead(II) acetate

39.   <u>Ternary Ionic Compound</u>   <u>Chemical Formula</u>
- (a)   manganese(II) acetate $Mn(C_2H_3O_2)_2$
- (b)   copper(II) chlorite $Cu(ClO_2)_2$
- (c)   tin(II) phosphate $Sn_3(PO_4)_2$
- (d)   iron(III) hypochlorite $Fe(ClO)_3$

41.   Ternary Ionic Compound   Chemical Formula
    (a) francium sulfate Fr_2SO_4
    (b) sodium sulfite Na_2SO_3

43.   Ternary Ionic Compound   Chemical Formula
    (a) radium chlorate $Ra(ClO_3)_2$
    (b) barium bromate $Ba(BrO_3)_2$

Section 7.7 *Binary Molecular Compounds*

45.   Binary Molecular Compound   Systematic Name
    (a) BrF   bromine monofluoride
    (b) CF_4 carbon tetrafluoride
    (c) I_2O_4 diiodine tetraoxide
    (d) Cl_2O_3 dichlorine trioxide

47.   Binary Molecular Compound   Chemical Formula
    (a) dinitrogen pentaoxide N_2O_5
    (b) carbon tetrachloride CCl_4
    (c) iodine monobromide   IBr
    (d) dihydrogen sulfide H_2S

Section 7.8 *Binary Acids*

49.   Binary Acid   Systematic Name
    (a) HBr(*aq*)   hydrobromic acid
    (b) HI(*aq*)   hydroiodic acid

Section 7.9 *Ternary Oxyacids*

51.   Ternary Oxyacid   Systematic Name
    (a) $HClO_2$(*aq*)   chlorous acid
    (b) H_3PO_4(*aq*)   phosphoric acid

53.   Ternary Oxyacid   Chemical Formula
    (a) acetic acid $HC_2H_3O_2$(*aq*)
    (b) nitric acid HNO_3(*aq*)

55.   Ternary Oxyacid   Chemical Formula
    (a) hypoiodous acid   HIO(*aq*)
    (b) hypobromous acid   HBrO(*aq*)

General Exercises

57.

	Substance	Ionic Charge
(a)	iron metal atoms	0
(b)	ferrous ions	2+
(c)	iron(III) ions	3+
(d)	iron compounds	0

(Note: The total ionic charge on a compound must be zero.)

59.

	Polyatomic Anion	Valence Electrons
(a)	$IO_4^{?-}$	$7 + 4(6) = 31$ e$^-$ (thus, 1–)
(b)	$SiO_3^{?-}$	$4 + 3(6) = 22$ e$^-$ (thus, 2–)

(Note: The polyatomic anion with a charge of 2– must be SiO_3^{2-} because the total of the valence electrons is an even number (24).

61.

Ions	fluoride ion	oxide ion	nitride ion
copper(II) ion	CuF_2 copper(II) fluoride	CuO copper(II) oxide	Cu_3N_2 copper(II) nitride
cobalt(II) ion	CoF_2 cobalt(II) fluoride	CoO cobalt(II) oxide	Co_3N_2 cobalt(II) nitride
lead(II) ion	PbF_2 lead(II) fluoride	PbO lead(II) oxide	Pb_3N_2 lead(II) nitride

63.

Ions	hydroxide ion	sulfate ion	phosphate ion
mercurous ion	$Hg_2(OH)_2$ mercurous hydroxide	Hg_2SO_4 mercurous sulfate	$(Hg_2)_3(PO_4)_2$ mercurous phosphate
ferrous ion	$Fe(OH)_2$ ferrous hydroxide	$FeSO_4$ ferrous sulfate	$Fe_3(PO_4)_2$ ferrous phosphate
stannous ion	$Sn(OH)_2$ stannous hydroxide	$SnSO_4$ stannous sulfate	$Sn_3(PO_4)_2$ stannous phosphate

65.

	Compound	Suffix Ending			Compound	Suffix Ending
(a)	Na_2S	–ide		(b)	$H_2S(aq)$	–ic acid

67.

	Compound	Suffix Ending			Compound	Suffix Ending
(a)	Na_2SO_3	–ite		(b)	$H_2SO_3(aq)$	–ous acid

69.

	Compound	Suffix Ending			Compound	Suffix Ending
(a)	Na_2SO_4	–ate		(b)	$H_2SO_4(aq)$	–ic acid

71.

	Chemical Name	Chemical Formula
(a)	dihydrogen oxide (water)	H_2O
(b)	sodium hypochlorite (bleach)	$NaClO$
(c)	sodium hydroxide (caustic soda)	$NaOH$
(d)	sodium bicarbonate (baking soda)	$NaHCO_3$

73.

	Binary Compound	Systematic Name
(a)	BF_3	boron trifluoride
(b)	$SiCl_4$	silicon tetrachloride
(c)	As_2O_5	diarsenic pentaoxide
(d)	Sb_2O_3	diantimony trioxide

75.

Given Formula	Predicted Formula
ZrO_2	TiO_2

77.

Given Formula	Predicted Formula
$LaCl_3$	$AcCl_3$

Chemical Reactions

Section 8.1 *Evidence for Chemical Reactions*

1. (a) The release of ammonia gas is evidence for a *chemical reaction*.
    (b) The flame-extinguishing gas is evidence for a *chemical reaction*.

3. (a) The insoluble precipitate is evidence for a *chemical reaction*.
    (b) The increased volume is evidence for a *physical change*.

5. (a) The fizzing is evidence of gas bubbles and a *chemical reaction*.
    (b) The yellow flame is the release of energy and a *chemical reaction*.

Section 8.2 *Writing Chemical Equations*

7. $Mg(s) + Br_2(l) \rightarrow MgBr_2(s)$

9. $Zn(HCO_3)_2(s) \rightarrow ZnCO_3(s) + CO_2(g) + H_2O(g)$

11. $Cd(s) + Co(NO_3)_2(aq) \rightarrow Cd(NO_3)_2(aq) + Co(s)$

13. $LiI(aq) + AgNO_3(aq) \rightarrow AgI(s) + LiNO_3(aq)$

15. $HNO_3(aq) + NaOH(aq) \rightarrow NaNO_3(aq) + HOH(l)$

Section 8.3 *Balancing Chemical Equations*

17. (a) $3 H_2(g) + N_2(g) \rightarrow 2 NH_3(g)$
    (b) $Al_2(CO_3)_3(s) \rightarrow Al_2O_3(s) + 3 CO_2(g)$
    (c) $Sr(s) + 2 H_2O(l) \rightarrow Sr(OH)_2(aq) + H_2(g)$
    (d) $K_2SO_4(aq) + Ba(OH)_2(aq) \rightarrow BaSO_4(s) + 2 KOH(aq)$
    (e) $2 H_3PO_4(aq) + 3 Mn(OH)_2(s) \rightarrow Mn_3(PO_4)_2(s) + 6 HOH(l)$

19. (a) $H_2CO_3(aq) + 2\,NH_4OH(aq) \;\rightarrow\; (NH_4)_2CO_3(aq) + 2\,HOH(l)$
    (b) $Hg_2(NO_3)_2(aq) + 2\,NaBr(aq) \;\rightarrow\; Hg_2Br_2(s) + 2\,NaNO_3(aq)$
    (c) $Mg(s) + 2\,HC_2H_3O_2(aq) \;\rightarrow\; Mg(C_2H_3O_2)_2(aq) + H_2(g)$
    (d) $2\,LiNO_3(s) \;\rightarrow\; 2\,LiNO_2(s) + O_2(g)$
    (e) $2\,Pb(s) + O_2(g) \;\rightarrow\; 2\,PbO(s)$

Section 8.4 *Classifying Chemical Reactions*

21. Refer to the chemical reactions in Exercise 17.
    (a) combination reaction
    (b) decomposition reaction
    (c) single-replacement reaction
    (d) double-replacement reaction
    (e) neutralization reaction

23. Refer to the chemical reactions in Exercise 19.
    (a) neutralization reaction
    (b) double-replacement reaction
    (c) single-replacement reaction
    (d) decomposition reaction
    (e) combination reaction

Section 8.5 *Combination Reactions*

General Form: $A + Z \rightarrow AZ$

25. metal + oxygen gas $\rightarrow$ metal oxide
    (a) $2\,Ni(s) + O_2(g) \;\rightarrow\; 2\,NiO(s)$
    (b) $4\,Fe(s) + 3\,O_2(g) \;\rightarrow\; 2\,Fe_2O_3(s)$

27. nonmetal + oxygen gas $\rightarrow$ nonmetal oxide
    (a) $2\,C(s) + O_2(g) \;\rightarrow\; 2\,CO(g)$
    (b) $4\,P(s) + 5\,O_2(g) \;\rightarrow\; 2\,P_2O_5(s)$

29. metal + nonmetal $\rightarrow$ ionic compound
    (a) $2\,Cu(s) + Cl_2(g) \;\rightarrow\; 2\,CuCl(s)$
    (b) $Co(s) + S(s) \;\rightarrow\; CoS(s)$

31. (a) $4\,Cr(s) + 3\,O_2(g) \;\rightarrow\; 2\,Cr_2O_3(s)$
    (b) $2\,Cr(s) + N_2(g) \;\rightarrow\; 2\,CrN(s)$

33. (a) $4\,Li + O_2 \;\rightarrow\; 2\,Li_2O$
    (b) $2\,Ca + O_2 \;\rightarrow\; 2\,CaO$

35. (a) $2\,Na + I_2 \;\rightarrow\; 2\,NaI$
    (b) $3\,Ba + N_2 \;\rightarrow\; Ba_3N_2$

Section 8.6 *Decomposition Reactions*

General Form: $AX \rightarrow A + X$

37. metal hydrogen carbonate $\rightarrow$ metal carbonate + water + carbon dioxide
    (a) $2\,KHCO_3(s) \rightarrow K_2CO_3(s) + H_2O(g) + CO_2(g)$
    (b) $Sr(HCO_3)_2(s) \rightarrow SrCO_3(s) + H_2O(g) + CO_2(g)$

39. metal carbonate $\rightarrow$ metal oxide + carbon dioxide
    (a) $Co_2(CO_3)_3(s) \rightarrow Co_2O_3(s) + 3\,CO_2(g)$
    (b) $Sn(CO_3)_2(s) \rightarrow SnO_2(s) + 2\,CO_2(g)$

41. oxygen-containing compounds $\rightarrow$ oxygen gas
    (a) $Ca(NO_3)_2(s) \rightarrow Ca(NO_2)_2(s) + O_2(g)$
    (b) $2\,Ag_2SO_4(s) \rightarrow 2\,Ag_2SO_3(s) + O_2(g)$

43. metal hydrogen carbonate $\rightarrow$ metal carbonate + water + carbon dioxide
    (a) $2\,AgHCO_3(s) \rightarrow Ag_2CO_3(s) + H_2O(g) + CO_2(g)$
    (b) $Zn(HCO_3)_2(s) \rightarrow ZnCO_3(s) + H_2O(g) + CO_2(g)$

45. oxygen-containing compounds $\rightarrow$ oxygen gas
    (a) $2\,NaClO_3(s) \rightarrow 2\,NaCl(s) + 3\,O_2(g)$
    (b) $Ca(NO_3)_2(s) \rightarrow Ca(NO_2)_2(s) + O_2(g)$

Section 8.7 *The Activity Series Concept*

47.

	Element	Solution	Observation
(a)	Ag	$Ni(NO_3)_2(aq)$	no reaction; Ni > Ag
(b)	Sn	$Ni(NO_3)_2(aq)$	no reaction; Ni > Sn
(c)	Co	$Ni(NO_3)_2(aq)$	reaction; Co > Ni
(d)	Mn	$Ni(NO_3)_2(aq)$	reaction; Mn > Ni

49.

	Element	Solution	Observation
(a)	Ni	$HCl(aq)$	reaction; Ni > (H)
(b)	Zn	$HCl(aq)$	reaction; Zn > (H)
(c)	Cu	$HCl(aq)$	no reaction; Cu < (H)
(d)	Al	$HCl(aq)$	reaction; Al > (H)

51.

	Element	Solution	Observation
(a)	Li	$H_2O(l)$	reaction; Li is an active metal
(b)	Mg	$H_2O(l)$	no reaction; Mg is not an active metal
(c)	Ca	$H_2O(l)$	reaction; Ca is an active metal
(d)	Al	$H_2O(l)$	no reaction; Al is not an active metal

Section 8.8 *Single-Replacement Reactions*

$$\text{General Form:} \quad A + BZ \rightarrow AZ + B$$
$$\text{or:} \quad A + BZ \rightarrow NR$$

53. metal_1 + aqueous $\text{solution}_1 \rightarrow \text{metal}_2$ + aqueous solution_2
(a) $Cu(s) + Al(NO_3)_3(aq) \rightarrow NR$
(b) $2\,Al(s) + 3\,Cu(NO_3)_2(aq) \rightarrow 3\,Cu(s) + 2\,Al(NO_3)_3(aq)$

55. metal_1 + aqueous $\text{solution}_1 \rightarrow \text{metal}_2$ + aqueous solution_2
(a) $Ni(s) + Pb(C_2H_3O_2)_2(aq) \rightarrow Pb(s) + Ni(C_2H_3O_2)_2(aq)$
(b) $Pb(s) + Ni(C_2H_3O_2)_2(aq) \rightarrow NR$

57.   metal + aqueous acid $\rightarrow$ aqueous solution + hydrogen gas
(a) $Mg(s) + 2\,HNO_3(aq) \rightarrow Mn(NO_3)_2(aq) + H_2(g)$
(b) $Mn(s) + 2\,HCl(aq) \rightarrow MgCl_2(s) + H_2(g)$

59.   metal + water $\rightarrow$ metal hydroxide + hydrogen gas
(a) $2\,Li(s) + 2\,H_2O(l) \rightarrow 2\,LiOH(aq) + H_2(g)$
(b) $Ba(s) + 2\,H_2O(l) \rightarrow Ba(OH)_2(aq) + H_2(g)$

61. metal_1 + aqueous $\text{solution}_1 \rightarrow \text{metal}_2$ + aqueous solution_2
(a) $Zn(s) + Pb(NO_3)_2(aq) \rightarrow Zn(NO_3)_2(aq) + Pb(s)$
(b) $Cd(s) + Fe(NO_3)_2(aq) \rightarrow NR$

63.   metal + aqueous acid $\rightarrow$ aqueous solution + hydrogen gas
(a) $Zn(s) + 2\,HNO_3(aq) \rightarrow Zn(NO_3)_2(aq) + H_2(g)$
(b) $Cd(s) + 2\,HNO_3(aq) \rightarrow Cd(NO_3)_2(aq) + H_2(g)$

65.   metal + water $\rightarrow$ metal hydroxide + hydrogen gas
(a) $2\,K(s) + 2\,H_2O(l) \rightarrow 2\,KOH(aq) + H_2(g)$
(b) $Ba(s) + 2\,H_2O(l) \rightarrow Ba(OH)_2(aq) + H_2(g)$

Section 8.9 *Solubility Rules*

67.   (a) $Al(NO_3)_3$ soluble   (b) Na_2SO_4 soluble
      (c) $Co(OH)_2$ insoluble   (d) $FePO_4$ insoluble

69.   (a) Hg_2Cl_2 insoluble   (b) $HgCl_2$ soluble
      (c) $AgBr$ insoluble   (d) PbI_2 insoluble

Section 8.10 *Double-Replacement Reactions*

General Form: $AX + BZ \rightarrow AZ + BX$

or: $AX + BZ \rightarrow NR$

71. aqueous solution$_1$ + aqueous solution$_2$ → precipitate + aqueous solution$_3$
    (a) $CrI_3(aq) + 3\,NaOH(aq) \rightarrow Cr(OH)_3(s) + 3\,NaI(aq)$
    (b) $NiSO_4(aq) + Hg_2(NO_3)_2(aq) \rightarrow Hg_2SO_4(s) + Ni(NO_3)_2(aq)$

73. aqueous solution$_1$ + aqueous solution$_2$ → precipitate + aqueous solution$_3$
    (a) $2\,AgC_2H_3O_2(aq) + SrI_2(aq) \rightarrow 2\,AgI(s) + Sr(C_2H_3O_2)_2(aq)$
    (b) $FeSO_4(aq) + Ca(OH)_2(aq) \rightarrow Fe(OH)_2(s) + CaSO_4(aq)$

Section 8.11 *Neutralization Reactions*

General Form: $HX + BOH \rightarrow BX + HOH$

75. aqueous acid + aqueous base → aqueous salt + water
    (a) $H_2CO_3(aq) + 2\,LiOH(aq) \rightarrow Li_2CO_3(aq) + 2\,HOH(l)$
    (b) $2\,HNO_3(aq) + Ca(OH)_2(aq) \rightarrow Ca(NO_3)_2(aq) + 2\,HOH(l)$

77. aqueous acid + aqueous base → aqueous salt + water
    (a) $HNO_3(aq) + Ba(OH)_2(aq) \rightarrow Ba(NO_3)_2(aq) + 2\,HOH(l)$
    (b) $H_2SO_4(aq) + 2\,NaOH(aq) \rightarrow Na_2SO_4(aq) + 2\,HOH(l)$

General Exercises

79. If a chemical equation is impossible to balance, it is most likely that the equation has a chemical formula with an incorrect subscript.

81. (a) $3\,Fe(s) + 4\,H_2O(g) \rightarrow Fe_3O_4(s) + 4\,H_2(g)$
    (b) $4\,FeS(s) + 7\,O_2(g) \rightarrow 2\,Fe_2O_3(s) + 4\,SO_2(g)$

83. (a) $F_2(g) + 2\,NaBr(aq) \rightarrow Br_2(l) + 2\,NaF(aq)$
    (b) $Sb_2S_3(s) + 6\,HCl(aq) \rightarrow 2\,SbCl_3(aq) + 3\,H_2S(g)$

85. (a) $CH_4(g) + 2\,O_2(g) \rightarrow CO_2(g) + 2\,H_2O(g)$
    (b) $C_3H_8(g) + 5\,O_2(g) \rightarrow 3\,CO_2(g) + 4\,H_2O(g)$

87. **Industrial Preparation of Iron:**

    $Fe_2O_3(l) + 3\,CO(g) \rightarrow 2\,Fe(l) + 3\,CO_2(g)$

89.   **Ostwald Process for Manufacturing Nitric Acid:**

(1) $4\,NH_3(g) + 5\,O_2(g) \rightarrow 4\,NO(g) + 6\,H_2O(g)$
(2) $2\,NO(g) + O_2(g) \rightarrow 2\,NO_2(g)$
(3) $3\,NO_2(g) + H_2O(l) \rightarrow 2\,HNO_3(aq) + NO(g)$

The Mole Concept

Section 9.1 *Avogadro's Number*

1.

	Element	Average Mass		Element	Average Mass
(a)	Be	9.01 amu	(b)	Br	79.90 amu
(c)	Sc	44.96 amu	(d)	Se	78.96 amu

3.

	Element	Mass		Element	Mass
(a)	Be	9.01 g	(b)	Br	79.90 g
(c)	Sc	44.96 g	(d)	Se	78.96 g

Section 9.2 *Mole Calculations I*

5. (a) 1 mol Li $= 6.02 \times 10^{23}$ atoms Li
   (b) 1 mol F_2 $= 6.02 \times 10^{23}$ molecules F_2

7. (a) 6.02×10^{23} formula units NaCl $= 1$ mol NaCl
   (b) 6.02×10^{23} molecules NH_3 $= 1$ mol NH_3

9. (a) $0.125 \text{ mol Zn} \times \dfrac{6.02 \times 10^{23} \text{ atoms Zn}}{1 \text{ mol Zn}} = 7.52 \times 10^{22}$ atoms Zn

   (b) $0.250 \text{ mol Cl}_2 \times \dfrac{6.02 \times 10^{23} \text{ molecules Cl}_2}{1 \text{ mol Cl}_2} = 1.51 \times 10^{23}$ molecules Cl_2

   (c) $0.675 \text{ mol ZnCl}_2 \times \dfrac{6.02 \times 10^{23} \text{ formula units ZnCl}_2}{1 \text{ mol ZnCl}_2}$

   $= 4.06 \times 10^{23}$ formula units $ZnCl_2$

11.   (a) 1.25×10^{22} ~~atoms Mn~~ $\times \dfrac{1 \text{ mol Mn}}{6.02 \times 10^{23} \text{ ~~atoms Mn~~}} = 0.0208 \text{ mol Mn}$

       (b) 2.50×10^{23} ~~molecules SO₃~~ $\times \dfrac{1 \text{ mol } SO_3}{6.02 \times 10^{23} \text{ ~~molecules SO₃~~}} = 0.415 \text{ mol } SO_3$

       (c) 6.75×10^{24} ~~formula units MnSO₄~~ $\times \dfrac{1 \text{ mol } MnSO_4}{6.02 \times 10^{23} \text{ ~~formula units MnSO₄~~}}$

$$= 11.2 \text{ mol } MnSO_4$$

Section 9.3 *Molar Mass*

13.

	Element	Molar Mass
(a)	Al	26.98 g/mol
(b)	Si	28.09 g/mol
(c)	O_2	2(16.00 g O) = 32.00 g/mol
(d)	S_8	8(32.07 g S) = 256.56 g/mol

15.

	Compound	Molar Mass
(a)	CaS	40.08 g Ca + 32.07 g S = 72.15 g/mol
(b)	$CaSO_4$	40.08 g Ca + 32.07 g S + 4(16.00 g O) = 136.15 g/mol
(c)	$Fe(C_2H_3O_2)_2$	55.85 g + 2(12.01 g + 12.01 g + 1.01 g + 1.01 g + 1.01 g + 16.00 g + 16.00 g) = 173.95 g/mol
(d)	$Fe_3(PO_4)_2$	3(55.85 g) + 2(30.97 g + 16.00 g + 16.00 g + 16.00 g + 16.00 g) = 357.49 g/mol

Section 9.4 *Mole Calculations II*

17.   (a)   MM of Hg = 200.59 g/mol

$$2.95 \times 10^{23} \text{ ~~atoms Hg~~} \times \dfrac{1 \text{ ~~mol Hg~~}}{6.02 \times 10^{23} \text{ ~~atoms Hg~~}} \times \dfrac{200.59 \text{ g Hg}}{1 \text{ ~~mol Hg~~}}$$

$$= 98.3 \text{ g Hg}$$

       (b)   MM of N_2 = 2(14.01 g N) = 28.02 g/mol

$$1.16 \times 10^{22} \text{ ~~molecules N₂~~} \times \dfrac{1 \text{ ~~mol N₂~~}}{6.02 \times 10^{23} \text{ ~~molecules N₂~~}} \times \dfrac{28.02 \text{ g } N_2}{1 \text{ ~~mol N₂~~}}$$

$$= 0.540 \text{ g } N_2$$

(c) MM of $BaCl_2$ = 137.33 g Ba + 2(35.45 g Cl) = 208.23 g/mol

$$5.05 \times 10^{21} \text{ formula units } BaCl_2 \times \frac{1 \text{ mol } BaCl_2}{6.02 \times 10^{23} \text{ formula units } BaCl_2}$$

$$\times \frac{208.23 \text{ g } BaCl_2}{1 \text{ mol } BaCl_2} = 1.75 \text{ g } BaCl_2$$

19. (a) MM of K = 39.10 g/mol

$$1.50 \text{ g K} \times \frac{1 \text{ mol K}}{39.10 \text{ g K}} \times \frac{6.02 \times 10^{23} \text{ atoms K}}{1 \text{ mol K}}$$

$$= 2.31 \times 10^{22} \text{ atoms K}$$

(b) MM of O_2 = 2(16.00 g O) = 32.00 g/mol

$$0.470 \text{ g } O_2 \times \frac{1 \text{ mol } O_2}{32.00 \text{ g } O_2} \times \frac{6.02 \times 10^{23} \text{ molecules } O_2}{1 \text{ mol } O_2}$$

$$= 8.84 \times 10^{21} \text{ molecules } O_2$$

(c) MM of $AgClO_3$ = 107.87 g Ag + 35.45 g Cl + 3(16.00 g O) = 191.32 g/mol

$$0.555 \text{ g } AgClO_3 \times \frac{1 \text{ mol } AgClO_3}{191.32 \text{ g } AgClO_3} \times \frac{6.02 \times 10^{23} \text{ formula units } AgClO_3}{1 \text{ mol } AgClO_3}$$

$$= 1.75 \times 10^{21} \text{ formula units } AgClO_3$$

21. (a) $\dfrac{9.01 \text{ g}}{1 \text{ mol Be}} \times \dfrac{1 \text{ mol Be}}{6.02 \times 10^{23} \text{ atoms}} = 1.50 \times 10^{-23} \text{ g/atom}$

(b) $\dfrac{22.99 \text{ g}}{1 \text{ mol Na}} \times \dfrac{1 \text{ mol Na}}{6.02 \times 10^{23} \text{ atoms}} = 3.82 \times 10^{-23} \text{ g/atom}$

(c) $\dfrac{58.93 \text{ g}}{1 \text{ mol Co}} \times \dfrac{1 \text{ mol Co}}{6.02 \times 10^{23} \text{ atoms}} = 9.79 \times 10^{-23} \text{ g/atom}$

(d) $\dfrac{74.92 \text{ g}}{1 \text{ mol As}} \times \dfrac{1 \text{ mol As}}{6.02 \times 10^{23} \text{ atoms}} = 1.24 \times 10^{-22} \text{ g/atom}$

Section 9.5 *Molar Volume*

23. Standard conditions: 0 °C and 1 atmosphere (1 atm)

25. (a) MM of Ar = 39.95 g/mol

Density of Ar (at STP): $\dfrac{39.95\ \text{g}}{1\ \text{mol}} \times \dfrac{1\ \text{mol}}{22.4\ \text{L}} = 1.78\ \text{g/L}$

(b) MM of Cl_2 = 2(35.45 g Cl) = 70.90 g/mol

Density of Cl_2 (at STP): $\dfrac{70.90\ \text{g}}{1\ \text{mol}} \times \dfrac{1\ \text{mol}}{22.4\ \text{L}} = 3.17\ \text{g/L}$

(c) MM of CH_4 = 12.01 g C + 4(1.01 g H) = 16.05 g/mol

Density of CH_4 (at STP): $\dfrac{16.05\ \text{g}}{1\ \text{mol}} \times \dfrac{1\ \text{mol}}{22.4\ \text{L}} = 0.717\ \text{g/L}$

(d) MM of C_2H_6 = 2(12.01 g C) + 6(1.01 g H) = 30.08 g/mol

Density of C_2H_6 (at STP): $\dfrac{30.08\ \text{g}}{1\ \text{mol}} \times \dfrac{1\ \text{mol}}{22.4\ \text{L}} = 1.34\ \text{g/L}$

27. (a) MM of oxygen: $\dfrac{1.43\ \text{g}}{1\ \text{L}} \times \dfrac{22.4\ \text{L}}{\text{mol}} = 32.0\ \text{g/mol}$

(b) MM of phosphine: $\dfrac{1.52\ \text{g}}{1\ \text{L}} \times \dfrac{22.4\ \text{L}}{\text{mol}} = 34.0\ \text{g/mol}$

(c) MM of nitrous oxide: $\dfrac{1.97\ \text{g}}{1\ \text{L}} \times \dfrac{22.4\ \text{L}}{\text{mol}} = 44.1\ \text{g/mol}$

(d) MM of Freon-12: $\dfrac{5.40\ \text{g}}{1\ \text{L}} \times \dfrac{22.4\ \text{L}}{\text{mol}} = 121\ \text{g/mol}$

29.

Gas	Molecules	Mass	Volume at STP
chlorine, Cl_2	6.02×10^{23}	70.90 g	22.4 L
hydrogen chloride, HCl	6.02×10^{23}	36.45 g	22.4 L
sulfur tetrachloride, SCl_4	6.02×10^{23}	173.87 g	22.4 L

Section 9.6 *Mole Calculations III*

31. (a) MM of He = 4.00 g/mol

$$0.250 \text{ g He} \times \frac{1 \text{ mol He}}{4.00 \text{ g He}} \times \frac{22.4 \text{ L He}}{1 \text{ mol He}} = 1.40 \text{ L He}$$

   (b) MM of N_2 = 2(14.01 g N) = 28.02 g/mol

$$5.05 \text{ g } N_2 \times \frac{1 \text{ mol } N_2}{28.02 \text{ g } N_2} \times \frac{22.4 \text{ L } N_2}{1 \text{ mol } N_2} = 4.04 \text{ L } N_2$$

33. (a) MM of H_2S = 2(1.01 g H) + 32.07 g S = 34.09 g/mol

$$1.05 \text{ L } H_2S \times \frac{1 \text{ mol } H_2S}{22.4 \text{ L } H_2S} \times \frac{34.09 \text{ g } H_2S}{1 \text{ mol } H_2S} = 1.60 \text{ g } H_2S$$

   (b) MM of N_2O_3 = 2(14.01 g N) + 3(16.00 g O) = 76.02 g/mol

$$5.33 \text{ L } N_2O_3 \times \frac{1 \text{ mol } N_2O_3}{22.4 \text{ L } N_2O_3} \times \frac{76.02 \text{ g } N_2O_3}{1 \text{ mol } N_2O_3} = 18.1 \text{ g } N_2O_3$$

35. (a) $100.0 \text{ mL } H_2 \times \dfrac{1 \text{ L } H_2}{1000 \text{ mL } H_2} \times \dfrac{1 \text{ mol } H_2}{22.4 \text{ L } H_2} \times \dfrac{6.02 \times 10^{23} \text{ molecules } H_2}{1 \text{ mol } H_2}$

$$= 2.69 \times 10^{21} \text{ molecules } H_2$$

   (b) $70.5 \text{ mL } NH_3 \times \dfrac{1 \text{ L } NH_3}{1000 \text{ mL } NH_3} \times \dfrac{1 \text{ mol } NH_3}{22.4 \text{ L } NH_3} \times \dfrac{6.02 \times 10^{23} \text{ molecules } NH_3}{1 \text{ mol } NH_3}$

$$= 1.89 \times 10^{21} \text{ molecules } NH_3$$

37.

Gas	Molecules	Mass	Volume at STP
hydrogen, H_2	**1.50×10^{23}**	0.503 g	5.58 L
ammonia, NH_3	1.50×10^{23}	**4.25 g**	5.58 L
methane, CH_4	1.50×10^{23}	4.00 g	**5.58 L**

Section 9.7 *Percent Composition*

39. The law of definite composition states that a compound always contains the same elements in the same proportion by mass. Thus, the percent hydrogen by mass in a kilogram of water is 11%.

41. MM of $C_{12}H_{22}O_{11}$ = 12(12.01 g C) + 22(1.01 g H) + 11(16.00 g O)

    = 144.1 g C + 22.2 g H + 176.0 g O

    = 342.3 g/mol

$$\frac{144.1 \text{ g C}}{342.3 \text{ g } C_{12}H_{22}O_{11}} \times 100\% = 42.10\% \text{ C}$$

$$\frac{22.2 \text{ g H}}{342.3 \text{ g } C_{12}H_{22}O_{11}} \times 100\% = 6.49\% \text{ H}$$

$$\frac{176.0 \text{ g O}}{342.3 \text{ g } C_{12}H_{22}O_{11}} \times 100\% = 51.42\% \text{ O}$$

43. MM of $C_5H_{11}NSO_2$

    = 5(12.01 g C) + 11(1.01 g H) + 14.01 g N + 32.07 g S + 2(16.00 g O)

    = 60.05 g C + 11.11 g H + 14.01 g N + 32.07 g S + 32.00 g O

    = 149.24 g/mol

$$\frac{60.05 \text{ g C}}{149.24 \text{ g } C_5H_{11}NSO_2} \times 100\% = 40.24\% \text{ C}$$

$$\frac{11.11 \text{ g H}}{149.24 \text{ g } C_5H_{11}NSO_2} \times 100\% = 7.444\% \text{ H}$$

$$\frac{14.01 \text{ g N}}{149.24 \text{ g } C_5H_{11}NSO_2} \times 100\% = 9.388\% \text{ N}$$

$$\frac{32.07 \text{ g S}}{149.24 \text{ g } C_5H_{11}NSO_2} \times 100\% = 21.49\% \text{ S}$$

$$\frac{32.00 \text{ g O}}{149.24 \text{ g } C_5H_{11}NSO_2} \times 100\% = 21.44\% \text{ O}$$

45. MM of $HgNa_2C_{20}H_8Br_2O_6$

$$= 200.59 \text{ g Hg} + 2(22.99 \text{ g Na}) + 20(12.01 \text{ g C}) + 8(1.01 \text{ g H}) +$$
$$2(79.90 \text{ g Br}) + 6(16.00 \text{ g O})$$

$$= 200.59 \text{ g Hg} + 45.98 \text{ g Na} + 240.20 \text{ g C} + 8.08 \text{ g H} + 159.80 \text{ g Br} + 96.00 \text{ g O}$$

$$= 750.65 \text{ g/mol}$$

$$\frac{200.59 \text{ g Hg}}{750.65 \text{ g HgNa}_2\text{C}_{20}\text{H}_8\text{Br}_2\text{O}_6} \times 100\% = 26.722\% \text{ Hg}$$

$$\frac{45.98 \text{ g Na}}{750.65 \text{ g HgNa}_2\text{C}_{20}\text{H}_8\text{Br}_2\text{O}_6} \times 100\% = 6.125\% \text{ Na}$$

$$\frac{240.20 \text{ g C}}{750.65 \text{ g HgNa}_2\text{C}_{20}\text{H}_8\text{Br}_2\text{O}_6} \times 100\% = 31.999\% \text{ C}$$

$$\frac{8.08 \text{ g H}}{750.65 \text{ g HgNa}_2\text{C}_{20}\text{H}_8\text{Br}_2\text{O}_6} \times 100\% = 1.08\% \text{ H}$$

$$\frac{159.80 \text{ g Br}}{750.65 \text{ g HgNa}_2\text{C}_{20}\text{H}_8\text{Br}_2\text{O}_6} \times 100\% = 21.288\% \text{ Br}$$

$$\frac{96.00 \text{ g O}}{750.65 \text{ g HgNa}_2\text{C}_{20}\text{H}_8\text{Br}_2\text{O}_6} \times 100\% = 12.79\% \text{ O}$$

Section 9.8 *Empirical Formula*

47. $0.500 \text{ g Sn} \times \dfrac{1 \text{ mol Sn}}{118.71 \text{ g Sn}} = 0.00421 \text{ mol Sn}$

$0.635 \text{ g Sn}_x\text{O}_y - 0.500 \text{ g Sn} = 0.135 \text{ g O}$

$0.135 \text{ g O} \times \dfrac{1 \text{ mol O}}{16.00 \text{ g O}} = 0.00844 \text{ mol O}$

$\text{Sn}_{\frac{0.00421}{0.00421}} \text{O}_{\frac{0.00844}{0.00421}} = \text{Sn}_{1.00}\text{O}_{2.00}$ The empirical formula is SnO_2.

49. $1.925 \text{ g Cu} \times \dfrac{1 \text{ mol Cu}}{63.55 \text{ g Cu}} = 0.03029 \text{ mol Cu}$

$2.410 \text{ g Cu}_x\text{O}_y - 1.925 \text{ g Cu} = 0.485 \text{ g O}$

$0.485 \text{ g O} \times \dfrac{1 \text{ mol O}}{16.00 \text{ g O}} = 0.0303 \text{ mol O}$

$\dfrac{\text{Cu}\,0.03029}{0.0303} \quad \dfrac{\text{O}\,0.0303}{0.0303} = \text{Cu}_{1.00}\text{O}_{1.00}$ The empirical formula is CuO.

51. $1.115 \text{ g Co} \times \dfrac{1 \text{ mol Co}}{58.93 \text{ g Co}} = 0.0189 \text{ mol Co}$

$2.025 \text{ g Co}_x\text{S}_y - 1.115 \text{ g Co} = 0.910 \text{ g S}$

$0.910 \text{ g S} \times \dfrac{1 \text{ mol S}}{32.07 \text{ g S}} = 0.0284 \text{ mol S}$

$\dfrac{\text{Co}\,0.0189}{0.0189} \quad \dfrac{\text{S}\,0.0284}{0.0189} = \text{Co}_{1.00}\text{S}_{1.50}$ The empirical formula is Co_2S_3.

53. $64.1 \text{ g Cu} \times \dfrac{1 \text{ mol Cu}}{63.55 \text{ g Cu}} = 1.01 \text{ mol Cu}$

$35.9 \text{ g Cl} \times \dfrac{1 \text{ mol Cl}}{35.45 \text{ g Cl}} = 1.01 \text{ mol Cl}$

$\dfrac{\text{Cu}\,1.01}{1.01} \quad \dfrac{\text{Cl}\,1.01}{1.01} = \text{Cu}_{1.00}\text{Cl}_{1.00}$ The empirical formula is CuCl.

55. $42.6 \text{ g Sn} \times \dfrac{1 \text{ mol Sn}}{118.71 \text{ g Sn}} = 0.359 \text{ mol Sn}$

$57.4 \text{ g Br} \times \dfrac{1 \text{ mol Br}}{79.90 \text{ g Br}} = 0.718 \text{ mol Br}$

$\dfrac{\text{Sn}\,0.359}{0.359} \quad \dfrac{\text{Br}\,0.718}{0.359} = \text{Sn}_{1.00}\text{Br}_{2.00}$ The empirical formula is SnBr_2.

57. $18.25 \text{ g C} \times \dfrac{1 \text{ mol C}}{12.01 \text{ g C}} = 1.52 \text{ mol C}$

$0.77 \text{ g H} \times \dfrac{1 \text{ mol H}}{1.01 \text{ g H}} = 0.76 \text{ mol H}$

$80.99 \text{ g Cl} \times \dfrac{1 \text{ mol Cl}}{35.45 \text{ g Cl}} = 2.28 \text{ mol Cl}$

$\dfrac{C\,1.52}{0.76} \quad \dfrac{H\,0.76}{0.76} \quad \dfrac{Cl\,2.28}{0.76} = C_{2.0}H_{1.0}Cl_{3.0}$ The empirical formula is C_2HCl_3.

Section 9.9 *Molecular Formula*

59.   MM of $C_9H_8O_4$ $= 9(12.01 \text{ g C}) + 8(1.01 \text{ g H}) + 4(16.00 \text{ g O})$
                           $= 108.19 \text{ g C} + 8.08 \text{ g H} + 64.00 \text{ g O}$
                           $= 180.17 \text{ g/mol}$

     Aspirin: $\dfrac{(C_9H_8O_4)_n}{C_9H_8O_4} = \dfrac{180 \text{ g/mol}}{180.17 \text{ g/mol}} \quad n \approx 1$

The molecular formula of aspirin is $(C_9H_8O_4)_1$ or $C_9H_8O_4$.

61.   MM of C_3H_8N $= 3(12.01 \text{ g C}) + 8(1.01 \text{ g H}) + 14.01 \text{ g N}$
                         $= 36.03 \text{ g C} + 8.08 \text{ g H} + 14.01 \text{ g N} = 58.12 \text{ g/mol}$

     Hexamethylene diamine: $\dfrac{(C_3H_8N)_n}{C_3H_8N} = \dfrac{115 \text{ g/mol}}{58.12 \text{ g/mol}} \quad n \approx 2$

The molecular formula of the compound is $(C_3H_8N)_2$ or $C_6H_{16}N_2$.

63.   <u>Empirical Formula</u>

$38.7 \text{ g C} \times \dfrac{1 \text{ mol C}}{12.01 \text{ g C}} = 3.22 \text{ mol C}$

$9.74 \text{ g H} \times \dfrac{1 \text{ mol H}}{1.01 \text{ g H}} = 9.64 \text{ mol H}$

$51.6 \text{ g O} \times \dfrac{1 \text{ mol O}}{16.00 \text{ g O}} = 3.23 \text{ mol O}$

$\dfrac{C\,3.22}{3.22} \quad \dfrac{H\,9.64}{3.22} \quad \dfrac{O\,3.23}{3.22} = C_{1.00}H_{2.99}O_{1.00}$ The empirical formula is CH_3O.

Molecular Formula

$$\text{MM of } CH_3O \quad = 12.01 \text{ g C} + 3(1.01 \text{ g H}) + 16.00 \text{ g O}$$
$$= 12.01 \text{ g C} + 3.03 \text{ g H} + 16.00 \text{ g O}$$
$$= 31.04 \text{ g/mol}$$

Ethylene glycol: $\dfrac{(CH_3O)_n}{CH_3O} = \dfrac{62 \text{ g/mol}}{31.04 \text{ g/mol}} \qquad n \approx 2$

The molecular formula of ethylene glycol is $(CH_3O)_2$ or $C_2H_6O_2$.

---

65.   Empirical Formula

$$24.8 \text{ g C} \times \frac{1 \text{ mol C}}{12.01 \text{ g C}} = 2.06 \text{ mol C}$$

$$2.08 \text{ g H} \times \frac{1 \text{ mol H}}{1.01 \text{ g H}} = 2.06 \text{ mol H}$$

$$73.1 \text{ g Cl} \times \frac{1 \text{ mol Cl}}{35.45 \text{ g Cl}} = 2.06 \text{ mol Cl}$$

$$\frac{C\ 2.06}{2.06} \frac{H\ 2.06}{2.06} \frac{Cl\ 2.06}{2.06} = C_{1.00}H_{1.00}Cl_{1.00} \qquad \text{The empirical formula is CHCl.}$$

Molecular Formula

$$\text{MM of CHCl} \quad = 12.01 \text{ g C} + 1.01 \text{ g H} + 35.45 \text{ g Cl}$$
$$= 48.47 \text{ g/mol}$$

Lindane: $\dfrac{(CHCl)_n}{CHCl} = \dfrac{290 \text{ g/mol}}{48.47 \text{ g/mol}} \qquad n \approx 6$

The molecular formula of lindane is $(CHCl)_6$ or $C_6H_6Cl_6$.

---

67.   Empirical Formula

$$74.0 \text{ g C} \times \frac{1 \text{ mol C}}{12.01 \text{ g C}} = 6.16 \text{ mol C}$$

$$8.70 \text{ g H} \times \frac{1 \text{ mol H}}{1.01 \text{ g H}} = 8.61 \text{ mol H}$$

$$17.3 \text{ g N} \times \frac{1 \text{ mol N}}{14.01 \text{ g N}} = 1.23 \text{ mol N}$$

$$\frac{C\ 6.16}{1.23} \frac{H\ 8.61}{1.23} \frac{N\ 1.23}{1.23} = C_{5.01}H_{7.00}N_{1.00} \qquad \text{The empirical formula is } C_5H_7N.$$

Molecular Formula

$$\text{MM of } C_5H_7N \quad = \quad 5(12.01 \text{ g C}) + 7(1.01 \text{ g H}) + 14.01 \text{ g N}$$
$$= \quad 60.05 \text{ g C} + 7.07 \text{ g H} + 14.01 \text{ g N}$$
$$= \quad 81.13 \text{ g/mol}$$

Nicotine: $\quad \dfrac{(C_5H_7N)_n}{C_5H_7N} \quad = \quad \dfrac{160 \text{ g/mol}}{81.13 \text{ g/mol}} \qquad n \approx 2$

The molecular formula of nicotine is $(C_5H_7N)_2$ or $C_{10}H_{14}N_2$.

General Exercises

69. 1 Einstein = 1 mole of photons = 6.02×10^{23} photons

71. $5 \text{ g Ni} \times \dfrac{1 \text{ mol Ni}}{58.69 \text{ g Ni}} \times \dfrac{6.02 \times 10^{23} \text{ atoms Ni}}{1 \text{ mol Ni}} = 5 \times 10^{22}$ atoms Ni

5×10^{22} atoms Ni $>>> 1 \times 10^{15}$ red blood cells

Thus, the number of Ni atoms in a 5-g nickel coin is 50,000,000 times greater than the number of red blood cells in 50,000 people!

73. $1 \text{ mol furry moles} \times \dfrac{6.02 \times 10^{23} \text{ furry moles}}{1 \text{ mol furry moles}} \times \dfrac{100 \text{ g}}{1 \text{ furry mole}} \times \dfrac{1 \text{ kg}}{1000 \text{ g}}$

$$= 6 \times 10^{22} \text{ kg}$$

6×10^{24} kg (Earth) $>> 6 \times 10^{22}$ kg (1 mol furry moles)
Thus, the Earth weighs about 100 times more than a mole of moles!

75. $0.500 \text{ g Ga} \times \dfrac{1 \text{ mol Ga}}{69.72 \text{ g Ga}} = 0.00717$ mol Ga

$0.672 \text{ g Ga}_xO_y - 0.500 \text{ g Ga} = 0.172 \text{ g O}$

$0.172 \text{ g O} \times \dfrac{1 \text{ mol O}}{16.00 \text{ g O}} = 0.0108$ mol O

$\dfrac{\text{Ga } 0.00717}{0.00717} \dfrac{\text{O } 0.0108}{0.00717} = \text{Ga}_{1.00}O_{1.51}$ The empirical formula is Ga_2O_3.

77. $1 \text{ molecule } H_2O \times \dfrac{1 \text{ mol } H_2O}{6.02 \times 10^{23} \text{ molecules } H_2O} \times \dfrac{18.02 \text{ g } H_2O}{1 \text{ mol } H_2O} \times \dfrac{1 \text{ cm}^3 H_2O}{1.00 \text{ g } H_2O}$

$$= 2.99 \times 10^{-23} \text{ cm}^3 H_2O$$

79.   MM of $C_{12}H_{22}O_{11}$ = 12(12.01 g C) + 22(1.01 g H) + 11(16.00 g O)
                                      = 144.12 g C + 22.22 g H + 176.00 g O
                                        = 342.34 g/mol

$$1.00 \text{ g } \cancel{C_{12}H_{22}O_{11}} \times \frac{1 \text{ } \cancel{\text{mol}}}{342.34 \text{ } \cancel{\text{g}}} \times \frac{12 \text{ } \cancel{\text{mol C}}}{1 \text{ } \cancel{\text{mol } C_{12}H_{22}O_{11}}} \times \frac{6.02 \times 10^{23} \text{ atoms C}}{1 \text{ } \cancel{\text{mol C}}}$$

$$= 2.11 \times 10^{22} \text{ atoms C}$$

81. $$1 \text{ } \cancel{\text{molecule vitamin K}} \times \frac{1 \text{ } \cancel{\text{mol vitamin K}}}{6.02 \times 10^{23} \text{ } \cancel{\text{molecules vitamin K}}} \times \frac{173 \text{ } \cancel{\text{g vitamin K}}}{1 \text{ } \cancel{\text{mol vitamin K}}}$$

$$\times \frac{76.3 \text{ } \cancel{\text{g C}}}{100 \text{ } \cancel{\text{g vitamin K}}} \times \frac{1 \text{ } \cancel{\text{mol C}}}{12.01 \text{ } \cancel{\text{g C}}} \times \frac{6.02 \times 10^{23} \text{ atoms C}}{1 \text{ } \cancel{\text{mol C}}} = 11 \text{ atoms C}$$

(Note: Since the number of atoms must be a whole number, the answer is
             11 atoms, not 11.0 atoms.)

83. $$\frac{1 \text{ atom Cu}}{0.0118 \text{ } \cancel{\text{nm}^3}} \times \left(\frac{1 \times 10^9 \text{ } \cancel{\text{nm}}}{1 \text{ } \cancel{\text{m}}}\right)^3 \times \left(\frac{1 \text{ } \cancel{\text{m}}}{100 \text{ } \cancel{\text{cm}}}\right)^3 \times \frac{1 \text{ } \cancel{\text{cm}^3}}{8.92 \text{ } \cancel{\text{g Cu}}} \times \frac{63.55 \text{ } \cancel{\text{g Cu}}}{1 \text{ mol Cu}}$$

$$= \frac{6.04 \times 10^{23} \text{ atoms Cu}}{1 \text{ mol Cu}}$$

Chemical Equation Calculations

CHAPTER

10

Section 10.1 *Interpreting a Chemical Equation*

1.　General Equation: $A + 2B \rightarrow 3C + 2D$

(a) $1 \, \cancel{mol \, A} \times \dfrac{3 \, mol \, C}{1 \, \cancel{mol \, A}} = 3 \, mol \, C$

(b) $2 \, \cancel{L \, D} \times \dfrac{2 \, L \, B}{2 \, \cancel{L \, D}} = 2 \, L \, B$

3.　General Equation: $2A + 3B \rightarrow C$
 (a) conservation of mass law: $1.00 \, g \, A + 1.50 \, g \, B = 2.50 \, g \, C$
 (b) conservation of mass law: $2.75 \, g \, C - 1.00 \, g \, A = 1.75 \, g \, B$

5.　(a)　　　　　 $2\,N_2(g) + O_2(g) \rightarrow 2\,N_2O(g)$

MM of N_2	$= 2(14.01 \, g \, N)$	$= 28.02 \, g/mol$
MM of O_2	$= 2(16.00 \, g \, O)$	$= 32.00 \, g/mol$
MM of N_2O	$= 2(14.01 \, g \, N) + 16.00 \, g \, O$	$= 44.02 \, g/mol$

$$2(28.02 \, g \, N_2) + 32.00 \, g \, O_2 \rightarrow 2(44.02 \, g \, N_2O)$$
$$56.04 \, g \, N_2 + 32.00 \, g \, O_2 \rightarrow 88.04 \, g \, N_2O$$
$$88.04 \, g \rightarrow 88.04 \, g$$

(b)　　　　　 $N_2(g) + 2\,O_2(g) \rightarrow N_2O_4(g)$

MM of N_2	$= 2(14.01 \, g \, N)$	$= 28.02 \, g/mol$
MM of O_2	$= 2(16.00 \, g \, O)$	$= 32.00 \, g/mol$
MM of N_2O_4	$= 2(14.01 \, g \, N) + 4(16.00 \, g \, O)$	$= 92.02 \, g/mol$

$$28.02 \, g \, N_2 + 2(32.00 \, g \, O_2) \rightarrow 92.02 \, g \, N_2O_4$$
$$28.02 \, g \, N_2 + 64.00 \, g \, O_2 \rightarrow 92.02 \, g \, N_2O_4$$
$$92.02 \, g \rightarrow 92.02 \, g$$

Section 10.2 *Mole–Mole Problems*

7. $2 H_2O_2(l) \quad \rightarrow \quad 2 H_2(g) + O_2(g)$

Moles of O_2 produced:

$$5.00 \text{ mol } H_2O_2 \times \frac{1 \text{ mol } O_2}{2 \text{ mol } H_2O_2} = 2.50 \text{ mol } O_2$$

9. $2 Fe(s) + 3 Cl_2(g) \quad \rightarrow \quad 2 FeCl_3(s)$

Moles of Cl_2 that react:

$$0.100 \text{ mol Fe} \times \frac{3 \text{ mol } Cl_2}{2 \text{ mol Fe}} = 0.150 \text{ mol } Cl_2$$

Moles of $FeCl_3$ produced:

$$0.100 \text{ mol Fe} \times \frac{2 \text{ mol } FeCl_3}{2 \text{ mol Fe}} = 0.100 \text{ mol } FeCl_3$$

11. $C_3H_8(g) + 5 O_2(g) \quad \rightarrow \quad 3 CO_2(g) + 4 H_2O(g)$

Moles of $C_3H_8(g)$ that react:

$$2.50 \text{ mol } O_2 \times \frac{1 \text{ mol } C_3H_8}{5 \text{ mol } O_2} = 0.500 \text{ mol } C_3H_8$$

Moles of $CO_2(g)$ produced:

$$2.50 \text{ mol } O_2 \times \frac{3 \text{ mol } CO_2}{5 \text{ mol } O_2} = 1.50 \text{ mol } CO_2$$

Section 10.3 *Types of Stoichiometry Problems*

13.	Given	Unknown	Type of Stoichiometry
	mass	volume	mass–volume problem

15.	Given	Unknown	Type of Stoichiometry
	volume	volume	volume–volume problem

17.	Given	Unknown	Type of Stoichiometry
	mass	mass	mass–mass problem

Section 10.4 *Mass–Mass Problems*

19. $Cu(s) + 2 AgNO_3(aq) \rightarrow Cu(NO_3)_2(aq) + 2 Ag(s)$

MM of Cu = 63.55 g/mol
MM of Ag = 107.87 g/mol

$$1.00 \text{ g Cu} \times \frac{1 \text{ mol Cu}}{63.55 \text{ g Cu}} \times \frac{2 \text{ mol Ag}}{1 \text{ mol Cu}} \times \frac{107.87 \text{ g Ag}}{1 \text{ mol Ag}} = 3.39 \text{ g Ag}$$

21. $2 Zn(s) + O_2(g) \rightarrow 2 ZnO(s)$

MM of Zn = 65.39 g/mol
MM of ZnO = 81.39 g/mol

$1.00 \text{ g Zn} \times \dfrac{1 \text{ mol Zn}}{65.39 \text{ g Zn}} \times \dfrac{2 \text{ mol ZnO}}{2 \text{ mol Zn}} \times \dfrac{81.39 \text{ g ZnO}}{1 \text{ mol ZnO}} = 1.24 \text{ g ZnO}$

23. $2 Bi(s) + 3 Cl_2(g) \rightarrow 2 BiCl_3(s)$

MM of Bi = 208.98 g/mol
MM of BiCl$_3$ = 315.33 g/mol

$3.45 \text{ g Bi} \times \dfrac{1 \text{ mol Bi}}{208.98 \text{ g Bi}} \times \dfrac{2 \text{ mol BiCl}_3}{2 \text{ mol Bi}} \times \dfrac{315.33 \text{ g BiCl}_3}{1 \text{ mol BiCl}_3} = 5.21 \text{ g BiCl}_3$

25. $2 Co(s) + 3 HgCl_2(aq) \rightarrow 2 CoCl_3(aq) + 3 Hg(l)$

MM of Co = 58.93 g/mol
MM of Hg = 200.59 g/mol

$1.25 \text{ g Co} \times \dfrac{1 \text{ mol Co}}{58.93 \text{ g Co}} \times \dfrac{3 \text{ mol Hg}}{2 \text{ mol Co}} \times \dfrac{200.59 \text{ g Hg}}{1 \text{ mol Hg}} = 6.38 \text{ g Hg}$

27. $2 Na_3PO_4(aq) + 3 Ca(OH)_2(aq) \rightarrow Ca_3(PO_4)_2(s) + 6 NaOH(aq)$

MM of Na$_3$PO$_4$ = 163.94 g/mol
MM of Ca$_3$(PO$_4$)$_2$ = 310.18 g/mol

$1.78 \text{ g Na}_3\text{PO}_4 \times \dfrac{1 \text{ mol Na}_3\text{PO}_4}{163.94 \text{ g Na}_3\text{PO}_4} \times \dfrac{1 \text{ mol Ca}_3(\text{PO}_4)_2}{2 \text{ mol Na}_3\text{PO}_4} \times \dfrac{310.18 \text{ g Ca}_3(\text{PO}_4)_2}{1 \text{ mol Ca}_3(\text{PO}_4)_2}$

$= 1.68 \text{ g Ca}_3(\text{PO}_4)_2$

Section 10.5 *Mass–Volume Problems*

29. $2 LiHCO_3(s) \rightarrow Li_2CO_3(s) + H_2O(g) + CO_2(g)$

MM of LiHCO$_3$ = 67.96 g/mol

$1.59 \text{ g LiHCO}_3 \times \dfrac{1 \text{ mol LiHCO}_3}{67.96 \text{ g LiHCO}_3} \times \dfrac{1 \text{ mol CO}_2}{2 \text{ mol LiHCO}_3} \times \dfrac{22.4 \text{ L CO}_2}{1 \text{ mol CO}_2}$

$\times \dfrac{1000 \text{ mL CO}_2}{1 \text{ L CO}_2} = 262 \text{ mL CO}_2$

31. $Ca(ClO_3)_2(s) \rightarrow CaCl_2(s) + 3\,O_2(g)$

MM of $Ca(ClO_3)_2$ = 206.98 g/mol

$$5.00 \;\cancel{g\; Ca(ClO_3)_2} \times \frac{1\; mol\; \cancel{Ca(ClO_3)_2}}{206.98 \;\cancel{g\; Ca(ClO_3)_2}} \times \frac{3\; \cancel{mol\; O_2}}{1\; mol\;\cancel{Ca(ClO_3)_2}} \times \frac{22.4\; \cancel{L\; O_2}}{1\; \cancel{mol\; O_2}}$$

$$\times \; \frac{1000\; mL\; O_2}{1\; \cancel{L\; O_2}} \;=\; 1620\; mL\; O_2$$

33. $Mg(s) + H_2SO_4(aq) \rightarrow MgSO_4(aq) + H_2(g)$

MM of Mg = 24.31 g/mol

$$225 \;\cancel{mL\; H_2} \times \frac{1\; \cancel{L\; H_2}}{1000\; \cancel{mL\; H_2}} \times \frac{1\; \cancel{mol\; H_2}}{22.4\; \cancel{L\; H_2}} \times \frac{1\; \cancel{mol\; Mg}}{1\; \cancel{mol\; H_2}} \times \frac{24.31\; g\; Mg}{1\; \cancel{mol\; Mg}}$$

$$=\; 0.244\; g\; Mg$$

35. $2\,Na(s) + 2\,H_2O(l) \rightarrow 2\,NaOH(aq) + H_2(g)$

MM of Na = 22.99 g/mol

$$75.0 \;\cancel{mL\; H_2} \times \frac{1\; \cancel{L\; H_2}}{1000\; \cancel{mL\; H_2}} \times \frac{1\; \cancel{mol\; H_2}}{22.4\; \cancel{L\; H_2}} \times \frac{2\; \cancel{mol\; Na}}{1\; \cancel{mol\; H_2}} \times \frac{22.99\; g\; Na}{1\; \cancel{mol\; Na}}$$

$$=\; 0.154\; g\; Na$$

Section 10.6 *Volume–Volume Problems*

37. $H_2(g) + I_2(g) \rightarrow 2\,HI(g)$

$$125 \;\cancel{mL\; H_2} \times \frac{1\; mL\; I_2}{1\; \cancel{mL\; H_2}} \;=\; 125\; mL\; I_2$$

39. $3\,H_2(g) + N_2(g) \rightarrow 2\,NH_3(g)$

$$45.0 \;\cancel{mL\; NH_3} \times \frac{1\; mL\; N_2}{2\; \cancel{mL\; NH_3}} \;=\; 22.5\; mL\; N_2$$

41. $2\,CO(g) + O_2(g) \rightarrow 2\,CO_2(g)$

$$10.0 \;\cancel{L\; CO} \times \frac{1\; L\; O_2}{2\; \cancel{L\; CO}} \;=\; 5.00\; L\; O_2$$

43. $2 \, Cl_2(g) \; + \; 3 \, O_2(g) \; \rightarrow \; 2 \, Cl_2O_3(g)$

$$1.75 \; \cancel{L \, Cl_2O_3} \; \times \; \frac{2 \, L \, Cl_2}{2 \, \cancel{L \, Cl_2O_3}} \; = \; 1.75 \, L \, Cl_2$$

$$1.75 \; \cancel{L \, Cl_2} \; \times \; \frac{1000 \, mL \, Cl_2}{1 \, \cancel{L \, Cl_2}} \; = \; 1750 \, mL \, Cl_2$$

45. $2 \, SO_2(g) \; + \; O_2(g) \; \rightarrow \; 2 \, SO_3(g)$

$$25.0 \; \cancel{L \, O_2} \; \times \; \frac{2 \, L \, SO_3}{1 \, \cancel{L \, O_2}} \; = \; 50.0 \, L \, SO_3$$

47. $2 \, N_2(g) \; + \; 5 \, O_2(g) \; \rightarrow \; 2 \, N_2O_5(g)$

$$500.0 \; \cancel{cm^3 \, N_2O_5} \; \times \; \frac{2 \, cm^3 \, N_2}{2 \, \cancel{cm^3 \, N_2O_5}} \; = \; 500.0 \, cm^3 \, N_2$$

Section 10.7 *The Limiting Reactant Concept*

49. $2 \, NO(g) \; + \; O_2(g) \; \rightarrow \; 2 \, NO_2(g)$

$$1.00 \; \cancel{mol \, NO} \; \times \; \frac{2 \, mol \, NO_2}{2 \, \cancel{mol \, NO}} \; = \; 1.00 \, mol \, NO_2$$

$$1.00 \; \cancel{mol \, O_2} \; \times \; \frac{2 \, mol \, NO_2}{1 \, \cancel{mol \, O_2}} \; = \; 2.00 \, mol \, NO_2$$

The limiting reactant is NO because it produces less product. This reaction produces 1.00 mol NO_2.

51. $N_2(g) \; + \; O_2(g) \; \rightarrow \; 2 \, NO(g)$

$$5.00 \; \cancel{mol \, N_2} \; \times \; \frac{2 \, mol \, NO}{1 \, \cancel{mol \, N_2}} \; = \; 10.0 \, mol \, NO$$

$$5.00 \; \cancel{mol \, O_2} \; \times \; \frac{2 \, mol \, NO}{1 \, \cancel{mol \, O_2}} \; = \; 10.0 \, mol \, NO$$

Neither N_2 or O_2 is a limiting reactant because each produces the same amount of product. This reaction produces 10.0 mol NO.

53. $2 H_2(g) + O_2(g) \rightarrow 2 H_2O(g)$

$$5.00 \;\cancel{\text{mol } H_2} \times \frac{2 \text{ mol } H_2O}{2 \;\cancel{\text{mol } H_2}} = 5.00 \text{ mol } H_2O$$

$$5.00 \;\cancel{\text{mol } O_2} \times \frac{2 \text{ mol } H_2O}{1 \;\cancel{\text{mol } O_2}} = 10.0 \text{ mol } H_2O$$

The limiting reactant is H_2 because it produces less product. This reaction produces 5.00 mol H_2O.

55. $2 C_2H_6(g) + 7 O_2(g) \rightarrow 4 CO_2(g) + 6 H_2O(g)$

$$1.00 \;\cancel{\text{mol } C_2H_6} \times \frac{6 \text{ mol } H_2O}{2 \;\cancel{\text{mol } C_2H_6}} = 3.00 \text{ mol } H_2O$$

$$5.00 \;\cancel{\text{mol } O_2} \times \frac{6 \text{ mol } H_2O}{7 \;\cancel{\text{mol } O_2}} = 4.29 \text{ mol } H_2O$$

The limiting reactant is C_2H_6 because it produces less water. This reaction produces 3.00 mol H_2O.

57. $Co(s) + S(s) \rightarrow CoS(s)$

(a) $1.00 \;\cancel{\text{mol } Co} \times \dfrac{1 \text{ mol } CoS}{1 \;\cancel{\text{mol } Co}} = 1.00 \text{ mol } CoS$

$1.00 \;\cancel{\text{mol } S} \times \dfrac{1 \text{ mol } CoS}{1 \;\cancel{\text{mol } S}} = 1.00 \text{ mol } CoS$

Neither Co or S is a limiting reactant because each produces the same amount of product. This reaction produces 1.00 mol CoS.

After reaction:
mol of Co	= 1.00 − 1.00	= 0.00 mol
mol of S	= 1.00 − 1.00	= 0.00 mol
mol of CoS	= 0.00 + 1.00	= 1.00 mol

(b) $2.00 \;\cancel{\text{mol } Co} \times \dfrac{1 \text{ mol } CoS}{1 \;\cancel{\text{mol } Co}} = 2.00 \text{ mol } CoS$

$3.00 \;\cancel{\text{mol } S} \times \dfrac{1 \text{ mol } CoS}{1 \;\cancel{\text{mol } S}} = 3.00 \text{ mol } CoS$

The limiting reactant is Co because it produces less product. This reaction produces 2.00 mol CoS.

<u>After reaction:</u>

mol of Co	$=$	$2.00 - 2.00$	$= 0.00$ mol
mol of S	$=$	$3.00 - 2.00$	$= 1.00$ mol
mol of CoS	$=$	$0.00 + 2.00$	$= 2.00$ mol

Section 10.8 *Limiting Reactant Problems*

59. $FeO(l) + Mg(l) \rightarrow Fe(l) + MgO(s)$

$$40.0 \; \cancel{g\;FeO} \times \frac{1 \; \cancel{mol\;FeO}}{71.85 \; \cancel{g\;FeO}} \times \frac{1 \; \cancel{mol\;Fe}}{1 \; \cancel{mol\;FeO}} \times \frac{55.85 \; g\;Fe}{1 \; \cancel{mol\;Fe}} = 31.1 \; g \; Fe$$

$$10.0 \; \cancel{g\;Mg} \times \frac{1 \; \cancel{mol\;Mg}}{24.31 \; \cancel{g\;Mg}} \times \frac{1 \; \cancel{mol\;Fe}}{1 \; \cancel{mol\;Mg}} \times \frac{55.85 \; g\;Fe}{1 \; \cancel{mol\;Fe}} = 23.0 \; g \; Fe$$

Therefore, Mg is the limiting reactant because it will be consumed before FeO. This reaction produces 23.0 g Fe.

61. $Fe_2O_3(l) + 2\,Al(l) \rightarrow 2\,Fe(l) + Al_2O_3(g)$

$$175 \; \cancel{g\;Fe_2O_3} \times \frac{1 \; \cancel{mol\;Fe_2O_3}}{159.70 \; \cancel{g\;Fe_2O_3}} \times \frac{2 \; \cancel{mol\;Fe}}{1 \; \cancel{mol\;Fe_2O_3}} \times \frac{55.85 \; g\;Fe}{\cancel{mol\;Fe}} = 122 \; g \; Fe$$

$$37.5 \; \cancel{g\;Al} \times \frac{1 \; \cancel{mol\;Al}}{26.98 \; \cancel{g\;Al}} \times \frac{2 \; \cancel{mol\;Fe}}{2 \; \cancel{mol\;Al}} \times \frac{55.85 \; g\;Fe}{1 \; \cancel{mol\;Fe}} = 77.6 \; g \; Fe$$

Therefore, Al is the limiting reactant because it will be consumed before Fe_2O_3. This reaction produces 77.6 g Fe.

63. $Mg(OH)_2(s) + H_2SO_4(l) \rightarrow MgSO_4(s) + 2\,H_2O(l)$

$$1.00 \; \cancel{g\;Mg(OH)_2} \times \frac{1 \; \cancel{mol\;Mg(OH)_2}}{58.33 \; \cancel{g\;Mg(OH)_2}} \times \frac{1 \; \cancel{mol\;MgSO_4}}{1 \; \cancel{mol\;Mg(OH)_2}} \times \frac{120.38 \; g\;MgSO_4}{1 \; \cancel{mol\;MgSO_4}}$$

$$= 2.06 \; g \; MgSO_4$$

$$0.605 \; \cancel{g\;H_2SO_4} \times \frac{1 \; \cancel{mol\;H_2SO_4}}{98.09 \; \cancel{g\;H_2SO_4}} \times \frac{1 \; \cancel{mol\;MgSO_4}}{1 \; \cancel{mol\;H_2SO_4}} \times \frac{120.38 \; g\;MgSO_4}{1 \; \cancel{mol\;MgSO_4}}$$

$$= 0.742 \; g \; MgSO_4$$

Therefore, H_2SO_4 is the limiting reactant because it will be consumed before $Mg(OH)_2$. This reaction produces 0.742 g $MgSO_4$.

65. $2 \, Al(OH)_3(s) \; + \; 3 \, H_2SO_4(l) \; \rightarrow \; Al_2(SO_4)_3(aq) \; + \; 6 \, H_2O(l)$

$$1.00 \text{ g } \cancel{Al(OH)_3} \; \times \; \frac{1 \cancel{\text{ mol Al(OH)}_3}}{78.01 \text{ g } \cancel{Al(OH)_3}} \; \times \; \frac{6 \cancel{\text{ mol H}_2\text{O}}}{2 \cancel{\text{ mol Al(OH)}_3}} \; \times \; \frac{18.02 \text{ g } H_2O}{1 \cancel{\text{ mol H}_2\text{O}}}$$

$$= \; 0.693 \text{ g } H_2O$$

$$3.00 \text{ g } \cancel{H_2SO_4} \; \times \; \frac{1 \cancel{\text{ mol H}_2\text{SO}_4}}{98.09 \text{ g } \cancel{H_2SO_4}} \; \times \; \frac{6 \cancel{\text{ mol H}_2\text{O}}}{3 \cancel{\text{ mol H}_2\text{SO}_4}} \; \times \; \frac{18.02 \text{ g } H_2O}{1 \cancel{\text{ mol H}_2\text{O}}}$$

$$= \; 1.10 \text{ g } H_2O$$

Therefore, $Al(OH)_3$ is the limiting reactant because it will be consumed before H_2SO_4. This reaction produces 0.693 g H_2O.

67. $N_2(g) \; + \; 2 \, O_2(g) \; \rightarrow \; 2 \, NO_2(g)$

$$45.0 \text{ mL } \cancel{N_2} \; \times \; \frac{2 \text{ mL } NO_2}{1 \text{ mL } \cancel{N_2}} \; = \; 90.0 \text{ mL } NO_2$$

$$95.0 \text{ mL } \cancel{O_2} \; \times \; \frac{2 \text{ mL } NO_2}{2 \text{ mL } \cancel{O_2}} \; = \; 95.0 \text{ mL } NO_2$$

Therefore, N_2 is the limiting reactant because it will be consumed before O_2. This reaction produces 90.0 mL NO_2.

69. $2 \, N_2(g) \; + \; 3 \, O_2(g) \; \rightarrow \; 2 \, N_2O_3(g)$

$$70.0 \text{ mL } \cancel{N_2} \; \times \; \frac{2 \text{ mL } N_2O_3}{2 \text{ mL } \cancel{N_2}} \; = \; 70.0 \text{ mL } N_2O_3$$

$$45.0 \text{ mL } \cancel{O_2} \; \times \; \frac{2 \text{ mL } N_2O_3}{3 \text{ mL } \cancel{O_2}} \; = \; 30.0 \text{ mL } N_2O_3$$

Therefore, O_2 is the limiting reactant because it will be consumed before N_2. This reaction produces 30.0 mL N_2O_3.

71. $2 \, SO_2(g) \; + \; O_2(g) \; \rightarrow \; 2 \, SO_3(g)$

$$3.00 \text{ L } \cancel{SO_2} \; \times \; \frac{2 \text{ L } SO_3}{2 \text{ L } \cancel{SO_2}} \; = \; 3.00 \text{ L } SO_3$$

$$1.25 \text{ L } \cancel{O_2} \; \times \; \frac{2 \text{ L } SO_3}{1 \text{ L } \cancel{O_2}} \; = \; 2.50 \text{ L } SO_3$$

Therefore, O_2 is the limiting reactant because it will be consumed before SO_2. This reaction produces 2.50 L SO_3.

73. $4\,HCl(g) \;+\; O_2(g) \;\rightarrow\; 2\,Cl_2(g) \;+\; 2\,H_2O(g)$

$$50.0 \; \cancel{L\;HCl} \;\times\; \frac{2\;L\;Cl_2}{4\;\cancel{L\;HCl}} \;=\; 25.0\;L\;Cl_2$$

$$10.0 \; \cancel{L\;O_2} \;\times\; \frac{2\;L\;Cl_2}{1\;\cancel{L\;O_2}} \;=\; 20.0\;L\;Cl_2$$

Therefore, O_2 is the limiting reactant because it will be consumed before HCl. This reaction produces 20.0 L Cl_2.

Section 10.9 *Percent Yield*

75. Actual yield: 12.5 g PbI_2; theoretical yield: 13.9 g PbI_2

$$\text{Percent yield:} \qquad \frac{12.5\;\cancel{g}}{13.9\;\cancel{g}} \;\times\; 100\% \;=\; 89.9\%$$

77. Actual yield: 1.29 g $NaNO_2$; theoretical yield: 1.22 g $NaNO_2$

$$\text{Percent yield:} \;\; \frac{1.29\;\cancel{g}}{1.22\;\cancel{g}} \;\times\; 100\% \;=\; 106\%$$

General Exercises

79. The units associated with molar mass are grams per mole (g/mol).

81. $(NH_4)_2Cr_2O_7(s) \;\rightarrow\; Cr_2O_3(s) \;+\; 4\,H_2O(l) \;+\; N_2(g)$

MM of $(NH_4)_2Cr_2O_7$ = 252.10 g/mol
MM of Cr_2O_3 = 152.00 g/mol

$$1.54 \; \cancel{g\;(NH_4)_2Cr_2O_7} \;\times\; \frac{1\;\cancel{mol\;(NH_4)_2Cr_2O_7}}{252.10\;\cancel{g\;(NH_4)_2Cr_2O_7}} \;\times\; \frac{1\;\cancel{mol\;Cr_2O_3}}{1\;\cancel{mol\;(NH_4)_2Cr_2O_7}}$$

$$\times\; \frac{152.00\;g\;Cr_2O_3}{1\;\cancel{mol\;Cr_2O_3}} \;=\; 0.929\;g\;Cr_2O_3$$

83. $3 MnO_2(l) + 4 Al(l) \rightarrow 3 Mn(l) + 2 Al_2O_3(s)$

MM of Mn = 54.94 g/mol
MM of Al = 26.98 g/mol

$$1.00 \text{ kg Mn} \times \frac{1000 \text{ g Mn}}{1 \text{ kg Mn}} \times \frac{1 \text{ mol Mn}}{54.94 \text{ g Mn}} \times \frac{4 \text{ mol Al}}{3 \text{ mol Mn}} \times \frac{26.98 \text{ g Al}}{1 \text{ mol Al}}$$

$$= 655 \text{ g Al}$$

85. $Sb_2S_3(s) + 6 HCl(aq) \rightarrow 2 SbCl_3(aq) + 3 H_2S(g)$

MM of Sb_2S_3 = 339.71 g/mol

$$3.00 \text{ g Sb}_2\text{S}_3 \times \frac{1 \text{ mol Sb}_2\text{S}_3}{339.71 \text{ g Sb}_2\text{S}_3} \times \frac{3 \text{ mol H}_2\text{S}}{1 \text{ mol Sb}_2\text{S}_3} \times \frac{22.4 \text{ L H}_2\text{S}}{1 \text{ mol H}_2\text{S}} = 0.593 \text{ L H}_2\text{S}$$

87. $2 H_2O(l) \rightarrow 2 H_2(g) + O_2(g)$

MM of H_2O = 18.02 g/mol

$$100.0 \text{ mL H}_2\text{O} \times \frac{1.00 \text{ g H}_2\text{O}}{1 \text{ mL H}_2\text{O}} \times \frac{1 \text{ mol H}_2\text{O}}{18.02 \text{ g H}_2\text{O}} \times \frac{2 \text{ mol H}_2}{2 \text{ mol H}_2\text{O}} \times \frac{22.4 \text{ L H}_2}{1 \text{ mol H}_2}$$

$$= 124 \text{ L H}_2$$

89. $C_3H_8(g) + 5 O_2(g) \rightarrow 3 CO_2(g) + 4 H_2O(g)$

MM of C_3H_8 = 44.11 g/mol
MM of H_2O = 18.02 g/mol

$$10.0 \text{ g C}_3\text{H}_8 \times \frac{1 \text{ mol C}_3\text{H}_8}{44.11 \text{ g C}_3\text{H}_8} \times \frac{4 \text{ mol H}_2\text{O}}{1 \text{ mol C}_3\text{H}_8} \times \frac{18.02 \text{ g H}_2\text{O}}{1 \text{ mol H}_2\text{O}} = 16.3 \text{ g H}_2\text{O}$$

91. The limiting reactant is gasoline; the excess reactant is oxygen.

The Gaseous State

Section 11.1 *Properties of Gases*

1.   The following are observed properties of gases:
     (a)   Gases have a variable shape.
     (b)   Gases expand uniformly.
     (c)   Gases do compress infinitely; they condense to a liquid.
     (d)   Gases mix with other gases.

Section 11.2 *Atmospheric Pressure*

3.
	Units	Standard Pressure
(a)	millimeters of mercury	760 mm Hg
(b)	centimeters of mercury	76 cm Hg

5.   (a) $5.00 \text{ atm} \times \dfrac{14.7 \text{ psi}}{1 \text{ atm}} = 73.5 \text{ psi}$

     (b) $5.00 \text{ atm} \times \dfrac{76 \text{ cm Hg}}{1 \text{ atm}} = 3.80 \times 10^2 \text{ cm Hg}$

7.   (a) $30.8 \text{ in. Hg} \times \dfrac{101 \text{ kPa}}{29.9 \text{ in. Hg}} = 104 \text{ kPa}$

     (b) $30.8 \text{ in. Hg} \times \dfrac{760 \text{ torr}}{29.9 \text{ in. Hg}} = 783 \text{ torr}$

Section 11.3 *Variables Affecting Gas Pressure*

9.

	Change	Observation	Explanation
(a)	volume increases	pressure decreases	molecules are farther apart and collide less frequently
(b)	temperature increases	pressure increases	molecules are moving faster and collide with higher frequency and more energy
(c)	moles of gas increase	pressure increases	more molecules have more collisions

11.

	Change	Observation
(a)	volume increases	pressure decreases
(b)	temperature increases	pressure increases
(c)	moles of gas decrease	pressure decreases

13. Increasing the volume of a gas causes gas molecules to be farther apart, and they collide with the container less frequently. Since there are fewer collisions, the pressure of the gas decreases.

Section 11.4 *Boyle's Law*

15. Pressure vs. Volume

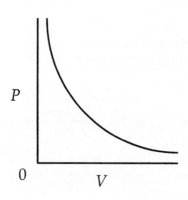

17. $P_1 \times V_{factor} = P_2$

$5.00 \text{ atm} \times \dfrac{1.75 \text{ L}}{2.50 \text{ L}} = 3.50 \text{ atm}$

19. $P_1 \times V_{factor} = P_2$

$5.00 \text{ L} \times \dfrac{1.55 \text{ atm}}{6.50 \text{ atm}} = 1.19 \text{ L}$

Section 11.5 *Charles's Law*

21.   Volume vs. Kelvin Temperature

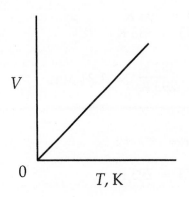

23. $V_1 \times T_{\text{factor}} = V_2$

25 °C + 273 = 298 K
50 °C + 273 = 323 K

$$555 \text{ mL} \times \frac{323 \cancel{K}}{298 \cancel{K}} = 602 \text{ mL}$$

25. $T_1 \times V_{\text{factor}} = T_2$

25 °C + 273 = 298 K

$$298 \text{ K} \times \frac{175 \cancel{\text{mL}}}{125 \cancel{\text{mL}}} = 417 \text{ K}$$

417 K − 273 = 144 °C

Section 11.6 *Gay-Lussac's Law*

27.   Pressure vs. Kelvin Temperature

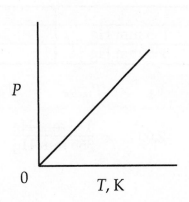

29. $P_1 \times T_{factor} = P_2$

$20\,°C + 273 = 293\,K$
$110\,°C + 273 = 383\,K$

$1.00\text{ atm} \times \dfrac{383\ \cancel{K}}{293\ \cancel{K}} = 1.31\text{ atm}$

31. $T_1 \times P_{factor} = T_2$

$20\,°C + 273 = 293\,K$

$293\,K \times \dfrac{0.750\ \cancel{atm}}{0.450\ \cancel{atm}} = 488\,K$

$488\,K - 273 = 215\,°C$

Section 11.7 Combined Gas Law

33.

	P	V	T
initial	5.00 atm	5.00 L	500 °C + 273 = 773 K
final	1.00 atm	V_2	273 K

$V_1 \quad\times\quad P_{factor} \quad\times\quad T_{factor} \quad=\quad V_2$

$5.00\text{ L} \times \dfrac{5.00\ \cancel{atm}}{1.00\ \cancel{atm}} \times \dfrac{273\ \cancel{K}}{773\ \cancel{K}} = 8.83\text{ L}$

35.

	P	V	T
initial	155 mm Hg	2.00 L	−50 °C + 273 = 223 K
final	365 mm Hg	V_2	75 °C + 273 = 348 K

$V_1 \quad\times\quad P_{factor} \quad\times\quad T_{factor} \quad=\quad V_2$

$2.00\text{ L} \times \dfrac{155\ \cancel{\text{mm Hg}}}{365\ \cancel{\text{mm Hg}}} \times \dfrac{348\ \cancel{K}}{223\ \cancel{K}} = 1.33\text{ L}$

37.

	P	V	T
initial	1.00 atm	0.750 L	273 K
final	P_2	0.100 L	25 °C + 273 = 298 K

$$P_1 \quad \times \quad V_{factor} \quad \times \quad T_{factor} \quad = \quad P_2$$

$$1.00 \text{ atm} \quad \times \quad \frac{0.750 \text{ L}}{0.100 \text{ L}} \times \frac{298 \text{ K}}{273 \text{ K}} = 8.19 \text{ atm}$$

39.

	P	V	T
initial	1.00 atm	1.00 L	273 K
final	2.00 atm	10.0 L	T_2

$$T_1 \quad \times \quad P_{factor} \quad \times \quad V_{factor} \quad = \quad T_2$$

$$273 \text{ K} \quad \times \quad \frac{2.00 \text{ atm}}{1.00 \text{ atm}} \quad \times \quad \frac{10.0 \text{ L}}{1.00 \text{ L}} \quad = \quad 5460 \text{ K}$$

$$5460 \text{ K} - 273 = 5187 \text{ °C} \ (5190 \text{ °C})$$

41.

	P	V	T
initial	225 mm Hg	0.500 L	−125 °C + 273 = 148 K
final	P_2	0.375 L	100 °C + 273 = 373 K

$$P_1 \quad \times \quad V_{factor} \quad \times \quad T_{factor} \quad = \quad P_2$$

$$225 \text{ mm Hg} \quad \times \quad \frac{0.500 \text{ L}}{0.375 \text{ L}} \times \frac{373 \text{ K}}{148 \text{ K}} = 756 \text{ mm Hg}$$

Section 11.8 *The Vapor Pressure Concept*

43. The vapor pressure of a liquid *increases* as the temperature *increases*.
Conversely, the vapor pressure *decreases* as the temperature *decreases*

45.

	Temperature	Vapor Pressure
(a)	25 °C	23.8 mm Hg
(b)	50 °C	92.5 mm Hg

Section 11.9 Dalton's Law

47. $P_{\text{nitrogen}} = 592 \text{ mm Hg}$
$P_{\text{oxygen}} = 160 \text{ mm Hg}$
$P_{\text{argon}} = 7 \text{ mm Hg}$
$P_{\text{trace gas}} = 1 \text{ mm Hg}$

$P_{\text{nitrogen}} + P_{\text{oxygen}} + P_{\text{argon}} + P_{\text{trace gas}} = P_{\text{atmosphere}}$
$592 \text{ mm Hg} + 160 \text{ mm Hg} + 7 \text{ mm Hg} + 1 \text{ mm Hg} = 760 \text{ mm Hg}$

49. $P_{\text{total}} = 1 \text{ atm} = 760 \text{ mm Hg}$
$P_{\text{sulfur dioxide}} = 150 \text{ mm Hg}$
$P_{\text{sulfur trioxide}} = 475 \text{ mm Hg}$

$P_{\text{sulfur dioxide}} + P_{\text{sulfur trioxide}} + P_{\text{oxygen}} = P_{\text{total}}$
$P_{\text{oxygen}} = P_{\text{total}} - P_{\text{sulfur dioxide}} - P_{\text{sulfur trioxide}}$
$P_{\text{oxygen}} = 760 \text{ mm Hg} - 150 \text{ mm Hg} - 475 \text{ mm Hg} = 135 \text{ mm Hg}$

51.   A gas may be collected over water to determine its volume. The volume of gas collected displaces an equal volume of water (see Figures 3.3 and 11.11). When collecting a gas over water to determine volume, we must assume that the gas is not very soluble in water.

53. $P_{\text{total}} = 766 \text{ mm Hg}$
$P_{\text{water vapor}} = 17.5 \text{ mm Hg}$

$P_{\text{water vapor}} + P_{\text{oxygen}} = P_{\text{total}}$
$P_{\text{oxygen}} = P_{\text{total}} - P_{\text{water vapor}}$
$P_{\text{oxygen}} = 766 \text{ mm Hg} - 17.5 \text{ mm Hg} = 749 \text{ mm Hg}$

Section 11.10 Ideal Gas Behavior

55.   The characteristics of an ideal gas are as follows:
   (1)   An ideal gas is mostly empty space and composed of tiny molecules, which have a negligible volume.
   (2)   An ideal gas has molecules that demonstrate rapid motion, move in straight lines, and travel in random directions.
   (3)   An ideal gas has molecules that show no attraction for one another.
   (4)   An ideal gas has molecules that demonstrate elastic collisions.
   (5)   In an ideal gas the average kinetic energy of molecules is proportional to the Kelvin temperature.

57.   A real gas behaves most like an ideal gas under the conditions of *high temperature* and *low pressure*.

59.
	Description	Gas
(a)	highest kinetic energy	all gases are equal
(b)	lowest kinetic energy	all gases are equal
(c)	highest velocity	He atoms
(d)	lowest velocity	Ar atoms

(Note: Since all three gases are at the same temperature, each has the same kinetic energy.)

61.   An ideal gas at 0 K exerts no pressure (0 mm Hg).

Section 11.11 *Ideal Gas Law*

63. $P = \dfrac{nRT}{V}$

$n = 0.500$ mol H_2

$T = 25\,°C + 273 = 298$ K

$V = 5.00$ L

$P = \dfrac{0.500\ \text{mol} \times 298\ \text{K}}{5.00\ \text{L}} \times \dfrac{0.0821\ \text{atm} \cdot \text{L}}{1\ \text{mol} \cdot \text{K}} = 2.45$ atm

65. $n = \dfrac{PV}{RT}$

$V = 10.0$ L

$T = 373$ K

$P = 125\ \text{psi} \times \dfrac{1\ \text{atm}}{14.7\ \text{psi}} = 8.50$ atm

$n = \dfrac{8.50\ \text{atm} \times 10.0\ \text{L}}{373\ \text{K}} \times \dfrac{1\ \text{mol} \cdot \text{K}}{0.0821\ \text{atm} \cdot \text{L}} = 2.78$ mol

General Exercises

67. $255\ \text{in.}^2 \times \dfrac{14.7\ \text{lb}}{1\ \text{in.}^2} = 37{,}500$ lb

69. $76.0\ \text{cm Hg} \times \dfrac{13.6\ \text{cm H}_2\text{O}}{1\ \text{cm Hg}} = 1030\ \text{cm H}_2\text{O}$

71. $P_{\text{total}} = 35\ \text{atm} + 125\ \text{atm} = 160$ atm

$P_{\text{total}} = 160\ \text{atm} \times \dfrac{14.7\ \text{psi}}{1\ \text{atm}} = 2350$ psi

73. $P_{\text{oxygen}} + P_{\text{water vapor}} = P_{\text{total}}$

$P_{\text{oxygen}} = P_{\text{total}} - P_{\text{water vapor}}$

$P_{\text{oxygen}} = 764 \text{ mm Hg} - 19.8 \text{ mm Hg} = 744 \text{ mm Hg}$

	P	V	T
initial	744 mm Hg	42.5 mL	22 °C + 273 = 295 K
final	760 mm Hg	V_2	273 K

$$42.5 \text{ mL} \times \frac{744 \text{ mm Hg}}{760 \text{ mm Hg}} \times \frac{273 \text{ K}}{295 \text{ K}} = 38.5 \text{ mL}$$

75. $P = \dfrac{nRT}{V}$

$$n = 1.51 \times 10^{23} \text{ molecules} \times \frac{1 \text{ mol}}{6.02 \times 10^{23} \text{ molecules}} = 0.251 \text{ mol}$$

$T = 25 \text{ °C} + 273 = 298 \text{ K}$ $\qquad V = 5.00 \text{ L}$

$$P = \frac{0.251 \text{ mol} \times 298 \text{ K}}{5.00 \text{ L}} \times \frac{0.0821 \text{ atm} \cdot \text{L}}{1 \text{ mol} \cdot \text{K}} = 1.23 \text{ atm}$$

77. Methane, CH_4, has a lower molecular mass than both ethane, C_2H_6, and hydrogen sulfide, H_2S, so CH_4 molecules have the fastest velocity.

79. $MM = \dfrac{gRT}{PV}$ $\qquad$ $g = 2.14 \text{ g}$
$\qquad\qquad\qquad\qquad\quad V = 1.00 \text{ L}$
$\qquad\qquad\qquad\qquad\quad P = 1 \text{ atm}$
$\qquad\qquad\qquad\qquad\quad T = 273 \text{ K}$

$$MM = \frac{2.14 \text{ g} \times 273 \text{ K}}{1 \text{ atm} \times 1.00 \text{ L}} \times \frac{0.0821 \text{ atm} \cdot \text{L}}{1 \text{ mol} \cdot \text{K}} = 48.0 \text{ g/mol} \quad (\text{thus, } \mathbf{O_3})$$

81. $MM = \dfrac{gRT}{PV}$ $\qquad$ $g = 1.95 \text{ g}$
$\qquad\qquad\qquad\qquad\quad V = 3.00 \text{ L}$
$\qquad\qquad\qquad\qquad\quad P = 1.25 \text{ atm}$
$\qquad\qquad\qquad\qquad\quad T = 20 \text{ °C} + 273 = 293 \text{ K}$

$$MM = \frac{1.95 \text{ g} \times 293 \text{ K}}{1.25 \text{ atm} \times 3.00 \text{ L}} \times \frac{0.0821 \text{ atm} \cdot \text{L}}{1 \text{ mol} \cdot \text{K}} = 12.5 \text{ g/mol}$$

83. $n = \dfrac{P\,V}{R\,T}$ $V = 1550 \text{ mL} \times \dfrac{1 \text{ L}}{1000 \text{ mL}} = 1.55 \text{ L}$

$P = 0.945 \text{ atm}$

$T = 50\,^{\circ}\text{C} + 273 = 323 \text{ K}$

$n = \dfrac{0.945 \text{ atm} \times 1.55 \text{ L}}{323 \text{ K}} \times \dfrac{1 \text{ mol} \cdot \text{K}}{0.0821 \text{ atm} \cdot \text{L}} = 0.0552 \text{ mol}$

$0.0552 \text{ mol } Cl_2 \times \dfrac{70.90 \text{ g } Cl_2}{1 \text{ mol } Cl_2} = 3.92 \text{ g } Cl_2$

85. $\dfrac{0.0821 \text{ atm} \cdot \text{L}}{1 \text{ mol} \cdot \text{K}} \times \dfrac{760 \text{ torr}}{1 \text{ atm}} = \dfrac{62.4 \text{ torr} \cdot \text{L}}{\text{mol} \cdot \text{K}}$

87. Deep-sea divers breathe a helium–oxygen mixture at depths below 200 feet. Since helium atoms have a lower molecular mass than air molecules (nitrogen and oxygen gases), helium atoms move faster across the vocal cords and produce a high-pitched voice.

Chemical Bonding

Section 12.1 *The Chemical Bond Concept*

1.   All Group IA/I elements, including K, have one valence electron. All Group VIIA/17 elements, including I, have seven valence electrons.

3.
Element	Atom	Ionic Bond
K	1 valence e⁻	0 valence e⁻
I	7 valence e⁻	8 valence e⁻

5.
	Compound	Bond			Compound	Bond
(a)	$AlCl_3$	ionic		(b)	H_2O	covalent
(c)	SO_3	covalent		(d)	$FeSO_4$	ionic

7.
	Substance	Particle			Substance	Particle
(a)	$HC_2H_3O_2$	molecule		(b)	ClO_2	molecule
(c)	$NaClO_3$	formula unit		(d)	TiO_2	formula unit

9.
	Substance	Particle			Substance	Particle
(a)	C_3H_8	molecule		(b)	Pt	atom
(c)	Fe_2O_3	formula unit		(d)	S_8	molecule

Section 12.2 *Ionic Bonds*

11.
	Ion	Ionic Charge			Ion	Ionic Charge
(a)	Li ion	1+ (Group 1)		(b)	Sr ion	2+ (Group 2)
(c)	Al ion	3+ (Group 13)		(d)	Pb ion	4+ (Group 14)

13.
	Ion	Ionic Charge			Ion	Ionic Charge
(a)	Cl ion	1– (Group 17)		(b)	I ion	1– (Group 17)
(c)	S ion	2– (Group 16)		(d)	P ion	3– (Group 15)

15.

	Ion	Electron Configuration
(a)	Li^+	$1s^2$
(b)	Al^{3+}	$1s^2\,2s^2\,2p^6$
(c)	Ca^{2+}	$1s^2\,2s^2\,2p^6\,3s^2\,3p^6$
(d)	Mg^{2+}	$1s^2\,2s^2\,2p^6$

17.

	Ion	Electron Configuration
(a)	Cl^-	$1s^2\,2s^2\,2p^6\,3s^2\,3p^6$
(b)	I^-	$1s^2\,2s^2\,2p^6\,3s^2\,3p^6\,4s^2\,3d^{10}\,4p^6\,5s^2\,4d^{10}\,5p^6$
(c)	S^{2-}	$1s^2\,2s^2\,2p^6\,3s^2\,3p^6$
(d)	P^{3-}	$1s^2\,2s^2\,2p^6\,3s^2\,3p^6$

19.

	Ion	Isoelectronic Noble Gas
(a)	S^{2-}	Ar
(b)	Cl^-	Ar
(c)	K^+	Ar
(d)	Ca^{2+}	Ar

21.

	Ion	Isoelectronic Noble Gas
(a)	Li^+	He
(b)	Al^{3+}	Ne
(c)	Ca^{2+}	Ar
(d)	Mg^{2+}	Ne

23.

	Ion	Isoelectronic Noble Gas
(a)	Cl^-	Ar
(b)	I^-	Xe
(c)	S^{2-}	Ar
(d)	P^{3-}	Ar

25. (a) Li atomic radius > Li ion radius
     (b) Mg atomic radius > Mg ion radius
     (c) F atomic radius < F ion radius
     (d) O atomic radius < O ion radius

27. The false statements are (b) and (c). The corrected statements are:
     (b) Cobalt atoms lose electrons, and sulfur atoms gain electrons.
     (c) The ionic radius of a cobalt ion is less than its atomic radius.

Section 12.3 *Covalent Bonds*

29. (a) The bond length in H–Cl is less than the sum of the atomic radii.
     (b) The bond length in O=O is less than the sum of the atomic radii.

31. The false statements are (a) and (d). The corrected statements are:
 (a) Valence electrons are *shared* between carbon and oxygen atoms.
 (d) Energy is *required* to break a C—O covalent bond.

Section 12.4 *Electron Dot Formulas of Molecules*

33.

Molecule	Valence Electrons	Electron Dot	Structural Formula
(a) H_2	$1 + 1 = 2$ e⁻	H:H	H—H
(b) F_2	$7 + 7 = 14$ e⁻	:F̈:F̈:	F — F
(c) HBr	$1 + 7 = 8$ e⁻	H:B̈r:	H—Br
(d) NH_3	$5 + 3(1) = 8$ e–	H:N̈:H with H	H—N—H with H

35.

Molecule	Valence Electrons	Electron Dot	Structural Formula
(a) HONO	$1 + 2(6) + 5 = 18$ e⁻	H:Ö:N::Ö	H—O—N=O
(b) SO_2	$6 + 2(6) = 18$ e⁻	:Ö:S::Ö	O—S=O
(c) C_2H_4	$2(4) + 4(1) = 12$ e⁻	H:C::C:H H H	H—C=C—H H H
(d) C_2H_2	$2(4) + 2(1) = 10$ e⁻	H:C:::C:H	H—C≡C—H

37.

Molecule	Valence Electrons	Electron Dot	Structural Formula
(a) CH_4	$4 + 4(1) = 8$ e⁻	H H : C : H H	H | H—C—H | H
(b) OF_2	$6 + 2(7) = 20$ e⁻	:F:O:F:	F—O—F
(c) H_2O_2	$2(1) + 2(6) = 14$ e⁻	H:O:O:H	H—O—O—H
(d) NF_3	$5 + 3(7) = 26$ e⁻	:F: :F:N:F:	F | F — N— F

Section 12.5 *Electron Dot Formulas of Polyatomic Ions*

39.

Polyatomic Ion	Valence Electrons	Electron Dot	Structural Formula
(a) BrO^-	$7 + 6 + 1 = 14$ e⁻	: Br : O: ⁻	[Br — O]⁻
(b) BrO_2^-	$7 + 2(6) + 1 = 20$ e⁻	:O: Br : O: ⁻	[O— Br — O]⁻
(c) BrO_3^-	$7 + 3(6) + 1 = 26$ e⁻	:O: Br :O: ⁻ :O:	[O— Br — O]⁻ | O
(d) BrO_4^-	$7 + 4(6) + 1 = 32$ e⁻	:O: :O: Br : O: ⁻ :O:	O | [O— Br — O]⁻ | O

41. | Polyatomic Ion | Valence Electrons | Electron Dot | Structural Formula |

(a) SO_4^{2-} $6 + 4(6) + 2 = 32$ e⁻

$$\begin{array}{c} :\!\overset{..}{\underset{..}{O}}\!: \\ :\!\overset{..}{O}\!:\!S\!:\!\overset{..}{O}\!: \\ :\!\overset{..}{\underset{..}{O}}\!: \end{array}{}^{2-}$$

$$\left[\begin{array}{c} O \\ | \\ O\!-\!S\!-\!O \\ | \\ O \end{array}\right]^{2-}$$

(b) HSO_4^- $1 + 6 + 4(6) + 1 = 32$ e⁻

$$\begin{array}{c} :\!\overset{..}{\underset{..}{O}}\!: \\ H\!:\!\overset{..}{O}\!:\!S\!:\!\overset{..}{O}\!: \\ :\!\overset{..}{\underset{..}{O}}\!: \end{array}{}^{-}$$

$$\left[\begin{array}{c} O \\ | \\ H\!-\!O\!-\!S\!-\!O \\ | \\ O \end{array}\right]^{-}$$

(c) SO_3^{2-} $6 + 3(6) + 2 = 26$ e⁻

$$\begin{array}{c} :\!\overset{..}{O}\!:\!S\!:\!\overset{..}{O}\!: \\ :\!\overset{..}{\underset{..}{O}}\!: \end{array}{}^{2-}$$

$$\left[\begin{array}{c} O\!-\!S\!-\!O \\ | \\ O \end{array}\right]^{2-}$$

(d) HSO_3^- $1 + 6 + 3(6) + 1 = 26$ e⁻

$$\begin{array}{c} H\!:\!\overset{..}{O}\!:\!\overset{..}{S}\!:\!\overset{..}{O}\!: \\ :\!\overset{..}{\underset{..}{O}}\!: \end{array}{}^{-}$$

$$\left[\begin{array}{c} H\!-\!O\!-\!S\!-\!O \\ | \\ O \end{array}\right]^{-}$$

43. | Polyatomic Ion | Valence Electrons | Electron Dot | Structural Formula |

(a) H_3O^+ $3(1) + 6 - 1 = 8$ e⁻

$$\begin{array}{c} H\!:\!\overset{..}{O}\!:\!H \\ | \\ H \end{array}{}^{+}$$

$$\left[\begin{array}{c} H\!-\!O\!-\!H \\ | \\ H \end{array}\right]^{+}$$

(b) OH^- $6 + 1 + 1 = 8$ e⁻

$$:\!\overset{..}{\underset{..}{O}}\!:\!H\;{}^{-}$$

$$[\,O\!-\!H\,]^{-}$$

(c) HS^- $1 + 6 + 1 = 8$ e⁻

$$H\!:\!\overset{..}{\underset{..}{S}}\!:\;{}^{-}$$

$$[\,H\!-\!S\,]^{-}$$

(d) CN^- $4 + 5 + 1 = 10$ e⁻

$$:\!C\!:\!:\!:\!N\!:\;{}^{-}$$

$$[\,C\!\equiv\!N\,]^{-}$$

Section 12.6 *Polar Covalent Bonds*

45.   Within a group of elements, the electronegativity increases from bottom to top in the periodic table.

47.   Nonmetals are more electronegative than metals.

49.

	More Electronegative		More Electronegative
(a)	**Cl** > Br	(b)	**O** > S
(c)	**Se** > As	(d)	N < **F**

(Note: The more electronegative element is in bold.)

51.

	Bond	Polarity		Bond	Polarity
(a)	Br—Cl	$3.0 - 2.8 = 0.2$	(b)	Br—F	$4.0 - 2.8 = 1.2$
(c)	I—Cl	$3.0 - 2.5 = 0.5$	(d)	I—Br	$2.8 - 2.5 = 0.3$

53.

	Polar Bonds Using Delta Notation		
(a)	δ^+ H—S δ^-	(b)	δ^- O—S δ^+
(c)	δ^+ N—F δ^-	(d)	δ^+ S—Cl δ^-

(Note: δ^- indicates the more electronegative atom and δ^+ indicates the more electropositive atom.)

Section 12.7 *Nonpolar Covalent Bonds*

55.

	Bond	Polarity	Classification
(a)	Cl—Cl	$3.0 - 3.0 = 0$	nonpolar
(b)	Cl—N	$3.0 - 3.0 = 0$	nonpolar
(c)	N—H	$3.0 - 2.1 = 0.9$	polar
(d)	H—P	$2.1 - 2.1 = 0$	nonpolar

Thus, (a), (b), and (d) are nonpolar.

57. H_2, N_2, Cl_2, and Br_2 occur naturally as diatomic molecules.

Section 12.8 *Coordinate Covalent Bonds*

(Note: Coordinate covalent bonds are indicated by a dash, —.)

59.

Molecule	Valence Electrons	Electron Dot	Coord. Cov. Bond
H**Br**O	$1 + 7 + 6 = 14$ e$^-$	H : Br : O:	H : Br — O:

61. | Molecule | Valence Electrons | Electron Dot | Coord. Cov. Bond |
|---|---|---|---|

$HBrO_2$ $1 + 7 + 2(6) = 20$ e⁻

$$H : \overset{..}{\underset{..}{Br}} : \overset{..}{\underset{..}{O}} :$$
$$: \overset{}{\underset{..}{O}} :$$

$$H : \overset{..}{\underset{..}{Br}} - \overset{..}{\underset{..}{O}} :$$
$$: \overset{}{\underset{..}{O}} :$$

63. | Polyatomic Ion | Valence Electrons | Electron Dot | Coord. Cov. Bond |
|---|---|---|---|

NH_4^+ $5 + 4(1) - 1 = 8$ e⁻

$$\begin{array}{c} H \\ \overset{..}{} \\ H : N : H \\ H \end{array}{}^{+}$$

$$\begin{array}{c} H \\ \overset{..}{} \\ H : N : H \\ | \\ H \end{array}{}^{+}$$

65. | Polyatomic Ion | Valence Electrons | Electron Dot | Coord. Cov. Bond |
|---|---|---|---|

NO_3^- $5 + 3(6) + 1 = 24$ e⁻

$$\begin{array}{c} : \overset{..}{O} : \\ : \overset{..}{O} : N :: \overset{..}{O} : \end{array}{}^{-}$$

$$\begin{array}{c} : \overset{..}{O} : \\ | \\ : \overset{..}{O} : N :: \overset{..}{O} : \end{array}{}^{-}$$

Section 12.9 *Hydrogen Bonds*

67.

$$\begin{array}{ccc} H & & H \\ | & & | \\ H-N: & \text{------} & H-N: \\ | & \uparrow & | \\ H & & H \end{array}$$

hydrogen bond

69. The bond energy is much less in a hydrogen bond than in a polar covalent bond.

Section 12.10 *Shapes of Molecules*

71. | Formula | Electron Pair | Molecular Shape | Bond Angle |
|---|---|---|---|
| SiH_4 | tetrahedral | tetrahedral | 109.5° |

73. | Formula | Electron Pair | Molecular Shape | Bond Angle |
|---|---|---|---|
| NI_3 | tetrahedral | trigonal pyramidal | 107° |

75.

Formula	Electron Pair	Molecular Shape	Bond Angle
$H_2\mathbf{S}$	tetrahedral	bent	104.5°

77. Each C–F bond is polar, but the symmetrical tetrahedral arrangement of the four bonds produces a nonpolar molecule.

General Exercises

79.

	Substance	Particle
(a)	U	atom
(b)	F_2	molecule
(c)	UF_6	formula unit
(d)	HF	molecule

81.

	Ions	Chemical Formula
(a)	$3 Sr^{2+}$ and $2 As^{3-}$	Sr_3As_2
(b)	$1 Ra^{2+}$ and $1 O^{2-}$	RaO
(c)	$2 Al^{3+}$ and $3 CO_3{}^{2-}$	$Al_2(CO_3)_3$
(d)	$1 Cd^{2+}$ and $2 OH^-$	$Cd(OH)_2$

83. The radius of a sodium ion is less than the radius of a sodium atom because the sodium ion has one less energy level ($3s$). In addition, the ion has one more proton than electron, which draws the electrons closer to the nucleus.

85.

Bond	Polarity
B—Cl	$3.0 - 2.0 = 1.0$ (polar)

87.

Bond	Polarity
H—P	$2.1 - 2.1 = 0$ (nonpolar)

89.

Polar Bond and Delta Notation	
	δ^+ Ge—Cl δ^-

91.

Molecule	Valence Electrons	Electron Dot	Structural Formula
SiH_4	$4 + 4(1) = 8$ e$^-$	H H:Si:H H	H | H—Si—H | H

93. | Polyatomic Ion | Valence Electrons | Electron Dot | Structural Formula |
| --- | --- | --- | --- |
| AsO_3^{3-} | $5 + 3(6) + 3 = 26\ e^-$ | :Ö:As:Ö: ³⁻
 :Ö: | $[O{-}As{-}O]^{3-}$
 \|
 O |

95. | Formula | Electron Pair | Molecular Shape | Bond Angle |
| --- | --- | --- | --- |
| CS_2 | linear | linear | 180° |

97. | Formula | Electron Pair | Molecular Shape | Bond Angle |
| --- | --- | --- | --- |
| CCl_2O | trigonal planar | trigonal planar | 120° |

99. | Molecule | Valence Electrons | Electron Dot | Structural Formula |
| --- | --- | --- | --- |
| XeO_2 | $8 + 2(6) = 20\ e^-$ | :Ö:Xe:Ö: | $O{-}Xe{-}O$ |

Liquids and Solids

Section 13.1 *Properties of Liquids*

1.   General Observed Properties of Liquids
   (a)   True, liquids have a variable shape.
   (b)   True, most liquids usually flow readily.
   (c)   False, liquids do not compress significantly with changes in temperature or pressure; however, they may undergo a change in physical state.
   (d)   True, liquids are about 1000 times *more dense* than gases.

3.
	Substance	Temperature	Physical State
(a)	Ne	–225 °C	gas
(b)	Ne	–255 °C	solid
(c)	Ar	–175 °C	gas
(d)	Ar	–200 °C	solid
(e)	Kr	–155 °C	liquid
(f)	Kr	–150 °C	gas

Section 13.2 *The Intermolecular Bond Concept*

5.
	Liquid	Intermolecular Attraction
(a)	C_6H_{14}	dispersion force
(b)	C_2H_5–F	dipole force
(c)	CH_3–OH	hydrogen bond
(d)	CH_3–O–CH_3	dipole force

7.
	Liquid	Higher Vapor Pressure
(a)	CH_3COOH or C_2H_5Cl	C_2H_5Cl (weaker attraction)
(b)	C_2H_5OH or CH_3OCH_3	CH_3OCH_3 (weaker attraction)

9.
	Liquid	Higher Viscosity
(a)	CH_3COOH or C_2H_5Cl	CH_3COOH (stronger attraction)
(b)	C_2H_5OH or CH_3OCH_3	C_2H_5OH (stronger attraction)

Section 13.3 *Vapor Pressure, Boiling Point, Viscosity, Surface Tension*

11.   Water molecules in the liquid state evaporate to the gaseous state and form water vapor. Simultaneously, some water molecules in the vapor condense back to the liquid. By definition, the vapor pressure of water is the total pressure exerted by the water molecules in the vapor when *the rate of evaporation* **is equal to** *the rate of condensation.*

13.   The viscosity of a liquid is a measure of its resistance to flow. Since the intermolecular attraction in water is strong, water has less tendency to flow than other liquids with molecules the same size.

15.   If the molecules in a liquid are weakly attracted:
       (a)   the vapor pressure is *high.*   (b)   the boiling point is *low.*
       (c)   the viscosity is *low.*   (d)   the surface tension is *low.*

17.   The relationship between the vapor pressure of a liquid and its temperature is: *As the temperature of a liquid increases, the vapor pressure* increases.

19.   Temperature   Vapor Pressure
       (a)   45 °C   ~180 mm Hg
       (b)   60 °C   ~330 mm Hg

21.   The temperature at which the vapor pressure is equal to the atmospheric pressure is defined as the boiling point. Thus, the boiling point of acetone is 56 °C because the vapor pressure at this temperature is 760 mm Hg.

Section 13.4 *Properties of Solids*

23.   General Observed Properties of Solids
       (1)   Solids have a fixed shape and a fixed volume.
       (2)   Solids are either crystalline or noncrystalline.
       (3)   Solids do not compress or expand significantly.
       (4)   Solids usually have a slightly higher density than their corresponding liquids (H_2O and NH_3 are exceptions).
       (5)   Solids do not mix by diffusion.

25.   Substance   Temperature   Physical State
       (a)   Ga   0 °C   solid
       (b)   Ga   100 °C   liquid

Section 13.5 *Crystalline Solids*

27. Three examples of crystalline solids are NaCl (ionic solid), Cl_2 (molecular solid), and Na (metallic solid).

29.
	Crystalline Solid	Type of Particles
(a)	ionic solid	ions
(b)	molecular solid	molecules
(c)	metallic solid	metal atoms

31.
	Crystalline Solid	Classification
(a)	Zn	metallic solid
(b)	ZnO	ionic solid
(c)	P_4	molecular solid
(d)	IBr	molecular solid

Section 13.6 *Changes of Physical State*

33. Heating Curve for Ethanol

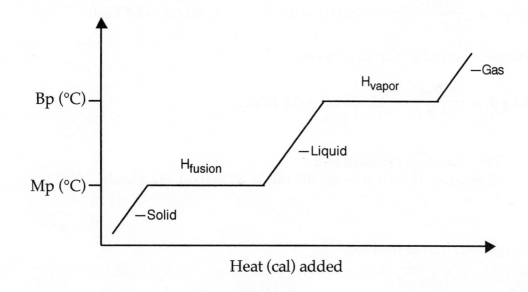

35. **Energy required to melt 50.0 g ice to water at 0 °C:**

$$50.0 \text{ g} \times \frac{80.0 \text{ cal}}{1 \text{ g}} = 4.00 \times 10^4 \text{ cal} \quad (40.0 \text{ kcal})$$

37. **Energy required to heat water:**

$$25.0 \text{ g} \times \frac{1.00 \text{ cal}}{1 \text{ g} \times {}^\circ\text{C}} \times (100.0 - 25.0) {}^\circ\text{C} = 1880 \text{ cal} \ (1.88 \text{ kcal})$$

Energy required to vaporize water:

$$25.0 \text{ g} \times \frac{540 \text{ cal}}{1 \text{ g}} = 13,500 \text{ cal} \ (13.5 \text{ kcal})$$

Total heat energy required:
1880 cal + 13,500 cal = 15,400 cal (15.4 kcal)

39. **Energy required to melt ice:**

$$115 \text{ g} \times \frac{80.0 \text{ cal}}{1 \text{ g}} = 9200 \text{ cal} \ (9.20 \text{ kcal})$$

Energy required to heat water:

$$115 \text{ g} \times \frac{1.00 \text{ cal}}{1 \text{ g} \times {}^\circ\text{C}} \times (100.0 - 0.0) {}^\circ\text{C} = 11,500 \text{ cal} \ (11.5 \text{ kcal})$$

Energy required to vaporize water:

$$115 \text{ g} \times \frac{540 \text{ cal}}{1 \text{ g}} = 62,100 \text{ cal} \ (62.1 \text{ kcal})$$

Total heat energy required:
9200 cal + 11,500 cal + 62,100 cal = 82,800 cal (82.8 kcal)

41. **Energy required to heat ice:**

$$38.5 \text{ g} \times \frac{0.50 \text{ cal}}{1 \text{ g} \times {}^\circ\text{C}} \times [0.0 - (-20.0)] {}^\circ\text{C} = 385 \text{ cal} \ (0.385 \text{ kcal})$$

Energy required to melt ice:

$$38.5 \text{ g} \times \frac{80.0 \text{ cal}}{1 \text{ g}} = 3080 \text{ cal} \ (3.08 \text{ kcal})$$

Energy required to heat water:

$$38.5 \cancel{g} \times \frac{1.00 \text{ cal}}{1 \cancel{g} \times \cancel{°C}} \times (100.0 - 0.0)\cancel{°C} = 3850 \text{ cal} \ (3.85 \text{ kcal})$$

Energy required to vaporize water:

$$38.5 \cancel{g} \times \frac{540 \text{ cal}}{1 \cancel{g}} = 20{,}800 \text{ cal} \ (20.8 \text{ kcal})$$

Total heat energy required:
385 cal + 3080 cal + 3850 cal + 20,800 cal = 28,100 cal (28.1 kcal)

43.   **Energy required to heat ice:**

$$100.0 \cancel{g} \times \frac{0.50 \text{ cal}}{1 \cancel{g} \times \cancel{°C}} \times [0.0 - (-40.0)]\cancel{°C} = 2000 \text{ cal} \ (2.00 \text{ kcal})$$

Energy required to melt ice:

$$100.0 \cancel{g} \times \frac{80.0 \text{ cal}}{1 \cancel{g}} = 8000 \text{ cal} \ (8.00 \text{ kcal})$$

Energy required to heat water:

$$100.0 \cancel{g} \times \frac{1.00 \text{ cal}}{1 \cancel{g} \times \cancel{°C}} \times (100.0 - 0.0)\cancel{°C} = 10{,}000 \text{ cal} \ (10.0 \text{ kcal})$$

Energy required to vaporize water:

$$100.0 \cancel{g} \times \frac{540 \text{ cal}}{1 \cancel{g}} = 54{,}000 \text{ cal} \ (54.0 \text{ kcal})$$

Energy required to heat steam:

$$100.0 \cancel{g} \times \frac{0.48 \text{ cal}}{1 \cancel{g} \times \cancel{°C}} \times (125.0 - 100.0)\cancel{°C} = 1200 \text{ cal} \ (1.20 \text{ kcal})$$

Total heat energy required:
2000 cal + 8000 cal + 10,000 cal + 54,000 cal + 1200 cal

$$= 75{,}200 \text{ cal} \ (75.2 \text{ kcal})$$

Section 13.7 *Structure of Water*

45.   In a water molecule, there are *two pairs* of bonding electrons and *two pairs* of nonbonding electrons.

47.   The observed bond angle in a water molecule is 104.5°.

49.   <u>Structural Formula of Water with Delta Notation</u>

51.   <u>Hydrogen Bond between HF molecules</u>

H—F ⎯⎯⎯ H—F

↑

hydrogen bond

Section 13.8 *Physical Properties of Water*

53.   As the water freezes to a solid, water molecules will form hydrogen bonds and the volume will expand. Thus, the cap will pop off, or the bottle will break.

55.   <u>Liquid</u>   <u>Higher Melting Point</u>
     (a) H_2O or H_2S H_2O (hydrogen bond attraction)
     (b) H_2S or H_2Se H_2Se (larger size, more attraction)

57.   <u>Liquid</u>   <u>Higher Heat of Fusion</u>
     (a) H_2O or H_2S H_2O (hydrogen bond attraction)
     (b) H_2S or H_2Se H_2Se (larger size, more attraction)

59.   <u>Property</u>   <u>As Molar Mass Increases</u>
     (a)   melting point   increases
     (b)   boiling point   increases
     (c)   heat of fusion   increases
     (d)   heat of vaporization   increases

Section 13.9 *Chemical Properties of Water*

61. $2 H_2(g) + O_2(g) \rightarrow 2 H_2O(l)$

63. (a) $2 Li(s) + 2 H_2O(l) \rightarrow 2 LiOH(aq) + H_2(g)$
    (b) $Na_2O(s) + H_2O(l) \rightarrow 2 NaOH(aq)$
    (c) $CO_2(g) + H_2O(l) \rightarrow H_2CO_3(aq)$

65. (a) $Mg(s) + H_2O(l) \rightarrow NR$
    (b) $SrO(s) + H_2O(l) \rightarrow Sr(OH)_2(aq)$
    (c) $N_2O_5(s) + H_2O(l) \rightarrow 2 HNO_3(aq)$

67. (a) $2 C_3H_6(g) + 9 O_2(g) \rightarrow 6 CO_2(g) + 6 H_2O(g)$
    (b) $Na_2Cr_2O_7 \bullet 2H_2O(s) \rightarrow Na_2Cr_2O_7(s) + 2 H_2O(g)$
    (c) $2 HF(aq) + Ca(OH)_2(aq) \rightarrow CaF_2(aq) + 2 HOH(l)$

69. (a) $2 C_4H_{10}(g) + 13 O_2(g) \rightarrow 8 CO_2(g) + 10 H_2O(g)$
    (b) $Co(C_2H_3O_2)_2 \bullet 4H_2O(s) \rightarrow Co(C_2H_3O_2)_2(s) + 4 H_2O(g)$
    (c) $2 HNO_3(aq) + Ba(OH)_2(aq) \rightarrow Ba(NO_3)_2(aq) + 2 HOH(l)$

Section 13.10 *Hydrates*

71.

	Chemical Formula	Systematic Name
(a)	$MgSO_4 \bullet 7H_2O$	magnesium sulfate heptahydrate
(b)	$Co(CN)_3 \bullet 3H_2O$	cobalt(III) cyanide trihydrate
(c)	$MnSO_4 \bullet H_2O$	manganese(II) sulfate monohydrate
(d)	$Na_2Cr_2O_7 \bullet 2H_2O$	sodium dichromate dihydrate

73.

	Systematic Name	Chemical Formula
(a)	sodium carbonate decahydrate	$Na_2CO_3 \bullet 10H_2O$
(b)	nickel(II) nitrate hexahydrate	$Ni(NO_3)_2 \bullet 6H_2O$
(c)	cobalt(III) iodide octahydrate	$CoI_3 \bullet 8H_2O$
(d)	chromium(III) acetate monohydrate	$Cr(C_2H_3O_2)_3 \bullet H_2O$

75. (a) Percentage of water in $SrCl_2 \bullet 6H_2O$

   MM of $SrCl_2 = 87.62 \text{ g} + 2(35.45 \text{ g}) = 158.52 \text{ g}$

   Percentage of water: $\dfrac{6(18.02 \text{ g})}{158.52 \text{ g} + 6(18.02 \text{ g})} \times 100\% = 40.55\% \text{ H}_2\text{O}$

(b) Percentage of water in $K_2Cr_2O_7 \cdot 2H_2O$

MM of $K_2Cr_2O_7$ = 2(39.10 g) + 2(52.00 g) + 7(16.00 g) = 294.20 g

Percentage of water: $\dfrac{2(18.02 \text{ g})}{294.20 \text{ g} + 2(18.02 \text{ g})}$ × 100% = 10.91% H_2O

(c) Percentage of water in $Co(CN)_3 \cdot 3H_2O$

MM of $Co(CN)_3$ = 58.93 g + 3(12.01 g) + 3(14.01 g) = 136.99 g

Percentage of water: $\dfrac{3(18.02 \text{ g})}{136.99 \text{ g} + 3(18.02 \text{ g})}$ × 100% = 28.30% H_2O

(d) Percentage of water in $Na_2CrO_4 \cdot 4H_2O$

MM of Na_2CrO_4 = 2(22.99 g) + 52.00 g + 4(16.00 g) = 161.98 g

Percentage of water: $\dfrac{4(18.02 \text{ g})}{161.98 \text{ g} + 4(18.02 \text{ g})}$ × 100% = 30.80% H_2O

77. (a) $NiCl_2 \cdot XH_2O(s)$ → $NiCl_2(s)$ + $XH_2O(g)$

21.7 g̶ H̶₂O̶ × $\dfrac{1 \text{ mol } H_2O}{18.02 \text{ g H}_2\text{O}}$ = 1.20 mol H_2O

78.3 g̶ N̶i̶C̶l̶₂ × $\dfrac{1 \text{ mol } NiCl_2}{129.59 \text{ g NiCl}_2}$ = 0.604 mol $NiCl_2$

$NiCl_2 \cdot \dfrac{1.20}{0.604} H_2O$ $\dfrac{1.20}{0.604}$ = 1.99 ≈ 2

Chemical Formula: $NiCl_2 \cdot 2H_2O$

(b) $Sr(NO_3)_2 \cdot XH_2O(s)$ $\rightarrow$ $Sr(NO_3)_2(s) + XH_2O(g)$

$$33.8 \text{ g } H_2O \times \frac{1 \text{ mol } H_2O}{18.02 \text{ g } H_2O} = 1.88 \text{ mol } H_2O$$

$$66.2 \text{ g } Sr(NO_3)_2 \times \frac{1 \text{ mol } Sr(NO_3)_2}{211.64 \text{ g } Sr(NO_3)_2} = 0.313 \text{ mol } Sr(NO_3)_2$$

$$Sr(NO_3)_2 \cdot \frac{1.88}{0.313} H_2O \qquad \frac{1.88}{0.313} = 6.01 \approx 6$$

Chemical Formula: $Sr(NO_3)_2 \cdot 6H_2O$

(c) $CrI_3 \cdot XH_2O(s)$ $\rightarrow$ $CrI_3(s) + XH_2O(g)$

$$27.2 \text{ g } H_2O \times \frac{1 \text{ mol } H_2O}{18.02 \text{ g } H_2O} = 1.51 \text{ mol } H_2O$$

$$72.8 \text{ g } CrI_3 \times \frac{1 \text{ mol } CrI_3}{432.70 \text{ g } CrI_3} = 0.168 \text{ mol } CrI_3$$

$$CrI_3 \cdot \frac{1.51}{0.168} H_2O \qquad \frac{1.51}{0.168} = 8.99 \approx 9$$

Chemical Formula: $CrI_3 \cdot 9H_2O$

(d) $Ca(NO_3)_2 \cdot XH_2O(s)$ $\rightarrow$ $Ca(NO_3)_2(s) + XH_2O(g)$

$$30.5 \text{ g } H_2O \times \frac{1 \text{ mol } H_2O}{18.02 \text{ g } H_2O} = 1.69 \text{ mol } H_2O$$

$$69.5 \text{ g } Ca(NO_3)_2 \times \frac{1 \text{ mol } Ca(NO_3)_2}{164.10 \text{ g } Ca(NO_3)_2} = 0.424 \text{ mol } Ca(NO_3)_2$$

$$Ca(NO_3)_2 \cdot \frac{1.69}{0.424} H_2O \qquad \frac{1.69}{0.424} = 3.99 \approx 4$$

Chemical Formula: $Ca(NO_3)_2 \cdot 4H_2O$

General Exercises

79.   Water covers about three-fourths (~75%) of Earth's surface.

81.   The vapor pressure of water is 650 mm Hg at ~95 °C. Therefore, the boiling point of water is ~95 °C at 650 torr.

83.   Sulfuric acid, H_2SO_4, is polar so we can predict that it behaves somewhat like water. That is, owing to surface tension sulfuric acid "raindrops" on Venus will form spheres similar to raindrops on Earth.

85.   **Energy released when ethylene glycol cools:**

$$1250 \text{ g} \times \frac{0.561 \text{ cal}}{1 \text{ g} \times \text{°C}} \times [25.0 - (-11.5)] \text{ °C} = 25{,}600 \text{ cal } (25.6 \text{ kcal})$$

Energy released when ethylene glycol solidifies:

$$1250 \text{ g} \times \frac{43.3 \text{ cal}}{1 \text{ g}} = 54{,}100 \text{ cal } (54.1 \text{ kcal})$$

Total heat energy released:

$$25{,}600 \text{ cal} + 54{,}100 \text{ cal} = 79{,}700 \text{ cal } (79.7 \text{ kcal})$$

Solutions

Section 14.1 *Gases in Solution*

1.
	Change	Result
(a)	temperature of solution *decreases*	solubility of O_2 gas *increases*
(b)	partial pressure of O_2 *decreases*	solubility of O_2 gas *decreases*

3.   standard solubility × pressure factor = new solubility

$$\frac{1.90 \text{ mL N}_2}{1 \text{ dL blood}} \times \frac{4.50 \text{ atm}}{1.00 \text{ atm}} = \frac{8.55 \text{ mL N}_2}{1 \text{ dL blood}}$$

5. $1.00 \text{ atm} \times \dfrac{100 \text{ g H}_2\text{O}}{0.12 \text{ g N}_2\text{O}} \times \dfrac{0.25 \text{ g N}_2\text{O}}{100 \text{ g H}_2\text{O}} = 2.1 \text{ atm}$

Section 14.2 *Liquids in Solution*

7.
(a)	polar solute + polar solvent = miscible
(b)	polar solute + nonpolar solvent = immiscible

9.
	Solvent	Classification
(a)	H_2O	polar
(b)	C_6H_{14}	nonpolar
(c)	C_3H_6O	polar
(d)	$CHCl_3$	nonpolar

11.
	Solvent	Miscible or Immiscible
(a)	C_2H_5OH	miscible with H_2O
(b)	C_7H_8	immiscible with H_2O
(c)	C_2HCl_3	immiscible with H_2O
(d)	$HC_2H_3O_2$	miscible with H_2O

13. Add several drops of the unknown liquid into a test tube containing water. If the unknown liquid is polar, it will be miscible with water. If the liquid is nonpolar, it will be immiscible and form two layers in the test tube.

Section 14.3 *Solids in Solution*

15. (a) polar solute + polar solvent = soluble
    (b) nonpolar solute + polar solvent = insoluble
    (c) ionic solute + polar solvent = soluble

17.

Compound	Soluble or Insoluble
(a) naphthalene, $C_{10}H_8$	insoluble in H_2O
(b) potassium hydroxide, KOH	soluble in H_2O
(c) calcium acetate, $Ca(C_2H_3O_2)_2$	soluble in H_2O
(d) trichlorotoluene, $C_7H_5Cl_3$	insoluble in H_2O
(e) glycine, $C_2H_5NO_2$	soluble in H_2O
(f) lactic acid, $HC_3H_5O_3$	soluble in H_2O

19.

Compound	Soluble or Insoluble
(a) cholesterol, $C_{27}H_{46}O$	fat soluble
(b) citric acid, $C_6H_8O_7$	water soluble
(c) fructose, $C_6H_{12}O_6$	water soluble
(d) stearic acid, $CH_3(CH_2)_{16}COOH$	fat soluble
(e) lactic acid, $CH_3CH(OH)COOH$	water soluble
(f) alanine, $CH_3CH(NH_2)COOH$	water soluble

Section 14.4 *The Dissolving Process*

21. Fructose molecule, $C_6H_{12}O_6$, dissolved in water:

$$H_2O$$
$$\vdots$$
$$H_2O \cdots\cdots C_6H_{12}O_6 \cdots\cdots H_2O$$
$$\vdots$$
$$H_2O$$

23. (a) Lithium bromide, LiBr, dissolved in water:

(b) Calcium chloride, $CaCl_2$, dissolved in water:

Section 14.5 *Rate of Dissolving*

25. The three factors that increase the rate of dissolving of a solid solute in a liquid solvent are:
   (1) heating the solution
   (2) stirring the solution
   (3) grinding the solid solute

Section 14.6 *Solubility and Temperature*

27. (a) ~35 g NaCl/100 g H_2O (b) ~35 g KCl/100 g H_2O

29. (a) ~36 g NaCl/100 g H_2O (b) ~41 g KCl/100 g H_2O

31. (a) ~0 °C   (b) ~70 °C

33. (a) ~100 °C   (b) ~35 °C

Section 14.7 *Unsaturated, Saturated, and Supersaturated Solutions*

35. (a) supersaturated at 25 °C
   (b) saturated at 55 °C
   (c) unsaturated at 75 °C

37. (a) supersaturated at 20 °C
   (b) saturated at 50 °C
   (c) unsaturated at 70 °C

39. $25.0 \text{ g } H_2O \times \dfrac{40.0 \text{ g rock salt}}{100 \text{ g } H_2O} = 10.0 \text{ g rock salt}$

   The solution is *saturated* at 30 °C.

41. (a) ~80 g solute remains in solution
   (b) ~20 g solute (100 g – 80 g) crystallizes from solution

Section 14.8 Mass/Mass Percent Concentration

43. $\dfrac{\text{mass of solute}}{\text{mass of solution}} \times 100\% = m/m\%$

(a) $\dfrac{20.0 \text{ g KI}}{20.0 \text{ g KI} + 100.0 \text{ g water}} \times 100\% = 16.7\%$

(b) $\dfrac{2.50 \text{ g AgC}_2\text{H}_3\text{O}_2}{2.50 \text{ g AgC}_2\text{H}_3\text{O}_2 + 95.0 \text{ g water}} \times 100\% = 2.56\%$

(c) $\dfrac{5.57 \text{ g SrCl}_2}{5.57 \text{ g SrCl}_2 + 225.0 \text{ g water}} \times 100\% = 2.42\%$

(d) $\dfrac{50.0 \text{ g sugar}}{50.0 \text{ g sugar} + 200.0 \text{ g water}} \times 100\% = 20.0\%$

45.

Solution	Unit Factors		
(a) 1.50% KBr	$\dfrac{1.50 \text{ g KBr}}{100 \text{ g solution}}$	and	$\dfrac{100 \text{ g solution}}{1.50 \text{ g KBr}}$
	$\dfrac{98.50 \text{ g H}_2\text{O}}{100 \text{ g solution}}$	and	$\dfrac{100 \text{ g solution}}{98.50 \text{ g H}_2\text{O}}$
	$\dfrac{1.50 \text{ g KBr}}{98.50 \text{ g H}_2\text{O}}$	and	$\dfrac{98.50 \text{ g H}_2\text{O}}{1.50 \text{ g KBr}}$
(b) 2.50% AlCl$_3$	$\dfrac{2.50 \text{ g AlCl}_3}{100 \text{ g solution}}$	and	$\dfrac{100 \text{ g solution}}{2.50 \text{ g AlCl}_3}$
	$\dfrac{97.50 \text{ g H}_2\text{O}}{100 \text{ g solution}}$	and	$\dfrac{100 \text{ g solution}}{97.50 \text{ g H}_2\text{O}}$
	$\dfrac{2.50 \text{ g AlCl}_3}{97.50 \text{ g H}_2\text{O}}$	and	$\dfrac{97.50 \text{ g H}_2\text{O}}{2.50 \text{ g AlCl}_3}$
(c) 3.75% AgNO$_3$	$\dfrac{3.75 \text{ g AgNO}_3}{100 \text{ g solution}}$	and	$\dfrac{100 \text{ g solution}}{3.75 \text{ g AgNO}_3}$
	$\dfrac{96.25 \text{ g H}_2\text{O}}{100 \text{ g solution}}$	and	$\dfrac{100 \text{ g solution}}{96.25 \text{ g H}_2\text{O}}$
	$\dfrac{3.75 \text{ g AgNO}_3}{96.25 \text{ g H}_2\text{O}}$	and	$\dfrac{96.25 \text{ g H}_2\text{O}}{3.75 \text{ g AgNO}_3}$

(d) 4.25% Li_2SO_4

$$\frac{4.25 \text{ g } Li_2SO_4}{100 \text{ g solution}} \quad \text{and} \quad \frac{100 \text{ g solution}}{4.25 \text{ g } Li_2SO_4}$$

$$\frac{95.75 \text{ g } H_2O}{100 \text{ g solution}} \quad \text{and} \quad \frac{100 \text{ g solution}}{95.75 \text{ g } H_2O}$$

$$\frac{4.25 \text{ g } Li_2SO_4}{95.75 \text{ g } H_2O} \quad \text{and} \quad \frac{95.75 \text{ g } H_2O}{4.25 \text{ g } Li_2SO_4}$$

47. (a) $5.36 \text{ g glucose} \times \dfrac{100 \text{ g solution}}{10.0 \text{ g glucose}} = 53.6 \text{ g solution}$

(b) $25.0 \text{ g sucrose} \times \dfrac{100 \text{ g solution}}{12.5 \text{ g sucrose}} = 200 \text{ g solution } (2.00 \times 10^2 \text{ g})$

49. (a) $85.0 \text{ g solution} \times \dfrac{2.00 \text{ g } FeBr_2}{100 \text{ g solution}} = 1.70 \text{ g } FeBr_2$

(b) $105.0 \text{ g solution} \times \dfrac{5.00 \text{ g } Na_2CO_3}{100 \text{ g solution}} = 5.25 \text{ g } Na_2CO_3$

51. (a) $100.0 \text{ g solution} \times \dfrac{95.0 \text{ g } H_2O}{100 \text{ g solution}} = 95.0 \text{ g } H_2O$

(b) $250.0 \text{ g solution} \times \dfrac{90.0 \text{ g } H_2O}{100 \text{ g solution}} = 225 \text{ g } H_2O$

Section 14.9 *Molar Concentration*

53. (a) MM of NaCl = 22.99 g/mol + 35.45 g/mol = 58.44 g/mol

100.0 mL solution = 0.1000 L solution

$$\frac{1.50 \text{ g NaCl}}{0.1000 \text{ L solution}} \times \frac{1 \text{ mol NaCl}}{58.45 \text{ g NaCl}} = 0.257 \text{ } M \text{ NaCl}$$

(b) MM of $K_2Cr_2O_7$ = 2(39.10 g/mol) + 2(52.00 g/mol) + 7(16.00 g/mol)

$$= 294.20 \text{ g/mol}$$

100.0 mL solution = 0.1000 L solution

$$\frac{1.50 \text{ g } K_2Cr_2O_7}{0.1000 \text{ L solution}} \times \frac{1 \text{ mol } K_2Cr_2O_7}{294.20 \text{ g } K_2Cr_2O_7} = 0.0510 \text{ } M \text{ } K_2Cr_2O_7$$

(c)   MM of $CaCl_2$ = 40.08 g/mol + 2(35.45 g/mol) = 110.98 g/mol

125 mL solution = 0.125 L solution

$$\frac{5.55 \text{ g } \cancel{CaCl_2}}{0.125 \text{ L solution}} \times \frac{1 \text{ mol } CaCl_2}{110.98 \text{ g } \cancel{CaCl_2}} = 0.400 \text{ } M \text{ } CaCl_2$$

(d)   MM of Na_2SO_4 = 2(22.99 g/mol) + 32.07 g/mol + 4(16.00 g/mol)

$$= 142.05 \text{ g/mol}$$

125 mL solution = 0.125 L solution

$$\frac{5.55 \text{ g } \cancel{Na_2SO_4}}{0.125 \text{ L solution}} \times \frac{1 \text{ mol } Na_2SO_4}{142.05 \text{ g } \cancel{Na_2SO_4}} = 0.313 \text{ } M \text{ } Na_2SO_4$$

55.

Solution	Unit Factors	
(a) 0.100 M LiI	$\dfrac{0.100 \text{ mol LiI}}{1 \text{ L solution}}$ and	$\dfrac{1 \text{ L solution}}{0.100 \text{ mol LiI}}$
	$\dfrac{0.100 \text{ mol LiI}}{1000 \text{ mL solution}}$ and	$\dfrac{1000 \text{ mL solution}}{0.100 \text{ mol LiI}}$
(b) 0.100 M $NaNO_3$	$\dfrac{0.100 \text{ mol } NaNO_3}{1 \text{ L solution}}$ and	$\dfrac{1 \text{ L solution}}{0.100 \text{ mol } NaNO_3}$
	$\dfrac{0.100 \text{ mol } NaNO_3}{1000 \text{ mL solution}}$ and	$\dfrac{1000 \text{ mL solution}}{0.100 \text{ mol } NaNO_3}$
(c) 0.500 M K_2CrO_4	$\dfrac{0.500 \text{ mol } K_2CrO_4}{1 \text{ L solution}}$ and	$\dfrac{1 \text{ L solution}}{0.500 \text{ mol } K_2CrO_4}$
	$\dfrac{0.500 \text{ mol } K_2CrO_4}{1000 \text{ mL solution}}$ and	$\dfrac{1000 \text{ mL solution}}{0.500 \text{ mol } K_2CrO_4}$
(d) 0.500 M $ZnSO_4$	$\dfrac{0.500 \text{ mol } ZnSO_4}{1 \text{ L solution}}$ and	$\dfrac{1 \text{ L solution}}{0.500 \text{ mol } ZnSO_4}$
	$\dfrac{0.500 \text{ mol } ZnSO_4}{1000 \text{ mL solution}}$ and	$\dfrac{1000 \text{ mL solution}}{0.500 \text{ mol } ZnSO_4}$

57. (a) MM of NaF = 22.99 g/mol + 19.00 g/mol = 41.99 g/mol

$$10.0 \; \cancel{\text{g NaF}} \times \frac{1 \; \cancel{\text{mol NaF}}}{41.99 \; \cancel{\text{g NaF}}} \times \frac{1 \; \text{L solution}}{0.275 \; \cancel{\text{mol NaF}}} = 0.866 \; \text{L solution}$$

(b) MM of $CdCl_2$ = 112.41 g/mol + 2(35.45 g/mol) = 183.31 g/mol

$$10.0 \; \cancel{\text{g CdCl}_2} \times \frac{1 \; \cancel{\text{mol CdCl}_2}}{183.31 \; \cancel{\text{g CdCl}_2}} \times \frac{1 \; \text{L solution}}{0.275 \; \cancel{\text{mol CdCl}_2}} = 0.198 \; \text{L solution}$$

(c) MM of K_2CO_3 = 2(39.10 g/mol) + 12.01 g/mol + 3(16.00 g/mol)

$$= 138.21 \; \text{g/mol}$$

$$10.0 \; \cancel{\text{g K}_2\text{CO}_3} \times \frac{1 \; \cancel{\text{mol K}_2\text{CO}_3}}{138.21 \; \cancel{\text{g K}_2\text{CO}_3}} \times \frac{1 \; \text{L solution}}{0.408 \; \cancel{\text{mol K}_2\text{CO}_3}} = 0.177 \; \text{L solution}$$

(d) MM of $Fe(ClO_3)_3$ = 55.85 g/mol + 3(35.45 g/mol) + 9(16.00 g/mol)

$$= 306.20 \; \text{g/mol}$$

$$10.0 \; \cancel{\text{g Fe(ClO}_3)_3} \times \frac{1 \; \cancel{\text{mol Fe(ClO}_3)_3}}{306.20 \; \cancel{\text{g Fe(ClO}_3)_3}} \times \frac{1 \; \text{L solution}}{0.408 \; \cancel{\text{mol Fe(ClO}_3)_3}}$$

$$= 0.0800 \; \text{L solution}$$

59. (a) MM of $FeCl_3$ = 55.85 g/mol + 3(35.45 g/mol) = 162.20 g/mol

$$2.25 \; \cancel{\text{L solution}} \times \frac{0.200 \; \cancel{\text{mol FeCl}_3}}{1 \; \cancel{\text{L solution}}} \times \frac{162.20 \; \text{g FeCl}_3}{1 \; \cancel{\text{mol FeCl}_3}} = 73.0 \; \text{g FeCl}_3$$

(b) MM of KIO_4 = 39.10 g/mol + 126.90 g/mol + 4(16.00 g/mol)

$$= 230.00 \; \text{g/mol}$$

$$2.25 \; \cancel{\text{L solution}} \times \frac{0.200 \; \cancel{\text{mol KIO}_4}}{1 \; \cancel{\text{L solution}}} \times \frac{230.00 \; \text{g KIO}_4}{1 \; \cancel{\text{mol KIO}_4}} = 104 \; \text{g KIO}_4$$

(c) MM of $ZnSO_4$ = 65.39 g/mol + 32.07 g/mol + 4(16.00 g/mol)

$$= 161.46 \text{ g/mol}$$

$$50.0 \text{ mL solution} \times \frac{0.295 \text{ mol ZnSO}_4}{1000 \text{ mL solution}} \times \frac{161.46 \text{ g ZnSO}_4}{1 \text{ mol ZnSO}_4}$$

$$= 2.38 \text{ g ZnSO}_4$$

(d) MM of $Ni(NO_3)_2$ = 58.69 g/mol + 2(14.01 g/mol) + 6(16.00 g/mol)

$$= 182.71 \text{ g/mol}$$

$$50.0 \text{ mL solution} \times \frac{0.295 \text{ mol Ni(NO}_3)_2}{1000 \text{ mL solution}} \times \frac{182.71 \text{ g Ni(NO}_3)_2}{1 \text{ mol Ni(NO}_3)_2}$$

$$= 2.69 \text{ g Ni(NO}_3)_2$$

61. MM of $CaSO_4$ = 40.08 g/mol + 32.07 g/mol + 4(16.00 g/mol) = 136.15 g/mol

100 mL solution = 0.100 L solution

$$\frac{0.209 \text{ g CaSO}_4}{0.100 \text{ L solution}} \times \frac{1 \text{ mol CaSO}_4}{136.15 \text{ g CaSO}_4} = 0.0154 \text{ M CaSO}_4$$

63. (a) $\dfrac{0.524 \text{ g glucose}}{10.483 \text{ g solution}} \times 100\% = 5.00\%$

(b) MM of $C_6H_{12}O_6$ = 6(12.01 g/mol) + 12(1.01 g/mol) + 6(16.00 g/mol)

$$= 180.18 \text{ g/mol}$$

10.0 mL solution = 0.0100 L solution

$$\frac{0.524 \text{ g C}_6\text{H}_{12}\text{O}_6}{0.0100 \text{ L solution}} \times \frac{1 \text{ mol C}_6\text{H}_{12}\text{O}_6}{180.18 \text{ g C}_6\text{H}_{12}\text{O}_6} = 0.291 \text{ M C}_6\text{H}_{12}\text{O}_6$$

Section 14.10 *Dilution of a Solution*

65. $10.0 \; \cancel{\text{mL solution}} \times \dfrac{12 \; \text{moL HCl}}{1000 \; \cancel{\text{mL solution}}} = 0.12 \; \text{mol HCl}$

$\dfrac{0.12 \; \text{mol HCl}}{250.0 \; \cancel{\text{mL solution}}} \times \dfrac{1000 \; \cancel{\text{mL solution}}}{1 \; \text{L solution}} = \dfrac{0.48 \; \text{mol HCl}}{1 \; \text{L solution}}$

$= 0.48 \; M \; \text{HCl}$

Algebraic Solution:

$$M_1 \times V_1 = M_2 \times V_2$$

$$\dfrac{12 \; M \times 10.0 \; \cancel{\text{mL}}}{250.0 \; \cancel{\text{mL}}} = 0.48 \; M \; \text{HCl}$$

67. $2.50 \; \cancel{\text{L solution}} \times \dfrac{0.10 \; \text{moL H}_2\text{SO}_4}{1 \; \cancel{\text{L solution}}} = 0.25 \; \text{mol H}_2\text{SO}_4$

$0.25 \; \cancel{\text{mol H}_2\text{SO}_4} \times \dfrac{1 \; \text{L solution}}{18 \; \cancel{\text{mol H}_2\text{SO}_4}} = 0.014 \; \text{L H}_2\text{SO}_4$

Algebraic Solution:

$$M_1 \times V_1 = M_2 \times V_2$$

$$\dfrac{0.10 \; M \times 2.50 \; \text{L}}{18 \; M} = 0.014 \; \text{L H}_2\text{SO}_4 \; (14 \; \text{mL H}_2\text{SO}_4)$$

Section 14.11 *Solution Stoichiometry*

69. $\text{MgBr}_2(aq) + 2 \; \text{AgNO}_3(aq) \rightarrow 2 \; \text{AgBr}(s) + \text{Mg(NO}_3)_2(aq)$

(a) $50.0 \; \cancel{\text{mL MgBr}_2} \times \dfrac{0.100 \; \cancel{\text{mol MgBr}_2}}{1000 \; \cancel{\text{mL MgBr}_2}} \times \dfrac{2 \; \text{mol AgNO}_3}{1 \; \cancel{\text{mol MgBr}_2}}$

$= 0.0100 \; \text{mol AgNO}_3$

$\dfrac{0.0100 \; \text{mol AgNO}_3}{13.9 \; \cancel{\text{mL solution}}} \times \dfrac{1000 \; \cancel{\text{mL solution}}}{1 \; \text{L solution}} = 0.719 \; M \; \text{AgNO}_3$

(b) $50.0 \; \cancel{\text{mL MgBr}_2} \times \dfrac{0.100 \; \cancel{\text{mol MgBr}_2}}{1000 \; \cancel{\text{mL MgBr}_2}} \times \dfrac{2 \; \cancel{\text{mol AgBr}}}{1 \; \cancel{\text{mol MgBr}_2}} \times \dfrac{187.77 \; \text{g AgBr}}{1 \; \cancel{\text{mol AgBr}}}$

$= 1.88 \; \text{g AgBr}$

General Exercises

71.   To demonstrate the Tyndall effect, colloidal-size particles must be present. The air in the auditorium most likely contains smoke and dust particles.

73.   A helium gas mixture is used rather than compressed air for two reasons:
       (1)   Helium is less soluble than nitrogen in blood.
       (2)   Helium atoms escape from blood more rapidly than nitrogen molecules; thus, allowing for faster decompression.

75.   100 mL solution = 0.100 L solution

$$1.00 \; \text{L solution} \times \frac{22.8 \text{ g SO}_2}{0.100 \text{ L solution}} = 228 \text{ g SO}_2$$

77.   Ethyl ether contains polar bonds to an oxygen atom, but it is only partially miscible in water. Ethyl ether has nonpolar characteristics because it has mostly nonpolar bonds between carbon and hydrogen atoms.

79.   The polar –OH in an alcohol molecule can form a hydrogen bond with water make it soluble. As the nonpolar C_xH_y- portion of the alcohol molecule increases in size, the alcohol becomes less soluble and eventually immiscible with water.

81.   (a)   In a 40% solution, ethanol is the solute and water is the solvent.
       (b)   In a 95% solution, water is the solute and ethanol is the solvent.

83.   MM of NaClO = 22.99 g/mol + 35.45 g/mol + 16.00 g/mol = 74.44 g/mol

$$\text{Solute: } 5.25 \; \text{g NaClO} \times \frac{1 \text{ mol NaClO}}{74.44 \text{ g NaClO}} = 0.0705 \text{ mol NaClO}$$

$$\text{Solution: } 100 \; \text{g bleach} \times \frac{1 \text{ mL bleach}}{1.04 \text{ g bleach}} \times \frac{1 \text{ L bleach}}{1000 \text{ mL bleach}} = 0.0962 \text{ L bleach}$$

$$\text{Molarity: } \frac{0.0705 \text{ mol NaClO}}{0.0962 \text{ L bleach}} = 0.733 \; M \; \text{NaClO}$$

CHAPTER
15

Acids and Bases

Section 15.1 *Properties of Acids and Bases*

1.　　　General Properties of an Acidic Solution
　　(a)　False, acidic solutions have a sour taste.
　　(b)　False, acidic solutions turn litmus paper red.
　　(c)　False, acidic solutions have a pH less than 7.
　　(d)　True, acidic solutions neutralize bases.

3.

	Aqueous Solution	pH	Classification
(a)	lime juice	2.7	weakly acidic
(b)	gastric juice	0.9	strongly acidic
(c)	distilled water	7.0	neutral
(d)	drain cleaner	13.5	strongly basic

Section 15.2 *Arrhenius Acids and Bases*

5.

	Acid	Ionization	Strength
(a)	$HClO_3(aq)$	~100%	strong
(b)	$HIO(aq)$	~1%	weak
(c)	$HBr(aq)$	~100%	strong
(d)	$HC_7H_5O_2(aq)$	~1%	weak

7.

	Formula	Classification		Formula	Classification
(a)	$HNO_3(aq)$	Arrhenius acid	(b)	$Mg(NO_3)_2(aq)$	salt
(c)	$Ca(OH)_2(aq)$	Arrhenius base	(d)	$H_2SO_3(aq)$	Arrhenius acid

9.   Arrhenius Acid   Arrhenius Base
     (a) HI(aq)   NaOH(aq)
     (b) HC$_2$H$_3$O$_2$(aq)   LiOH(aq)

11. Salt   Acid   Base
     (a) NaF(aq)   HF   NaOH
     (b) MgI$_2$(aq)   HI   Mg(OH)$_2$
     (c) Ca(NO$_3$)$_2$(aq) HNO$_3$ Ca(OH)$_2$
     (d) Li$_2$CO$_3$(aq)   H$_2$CO$_3$ LiOH

     Neutralization Reactions:

     (a) HF(aq) + NaOH(aq) → NaF(aq) + H$_2$O(l)
     (b) 2 HI(aq) + Mg(OH)$_2$(s) → MgI$_2$(aq) + 2 H$_2$O(l)
     (c) 2 HNO$_3$(aq) + Ca(OH)$_2$(aq) → Ca(NO$_3$)$_2$(aq) + 2 H$_2$O(l)
     (d) H$_2$CO$_3$(aq) + 2 LiOH(aq) → Li$_2$CO$_3$(aq) + 2 H$_2$O(l)

13. (a) 2 HNO$_3$(aq) + Ca(OH)$_2$(aq) → Ca(NO$_3$)$_2$(aq) + 2 H$_2$O(l)
     (b) H$_2$CO$_3$(aq) + Ba(OH)$_2$(aq) → BaCO$_3$(aq) + 2 H$_2$O(l)

Section 15.3 *Brønsted–Lowry Acids and Bases*

15. Acid   Base   Acid   Base
     (a) HC$_2$H$_3$O$_2$(aq) LiOH(aq)   (b) H$_2$SO$_4$(aq) NH$_3$(aq)

17. Acid   Base   Acid   Base
     (a) HI(aq)   H$_2$O(l)   (b) HC$_2$H$_3$O$_2$(aq) HS$^-$(aq)

19. (a) HF(aq) + NaHS(aq) → H$_2$S(aq) + NaF(aq)
     (b) HNO$_2$(aq) + NaC$_2$H$_3$O$_2$(aq) → NaNO$_2$(aq) + HC$_2$H$_3$O$_2$(aq)

Section 15.4 *Acid–Base Indicators*

21.   pH   Methyl Red Color   pH   Methyl Red Color
     (a) 4   red   (b) 8   yellow

23.   pH   Bromthymol Blue Color   pH   Bromthymol Blue Color
     (a) 5   yellow   (b) 9   blue

25.   pH   Phenolphthalein Color   pH   Phenolphthalein Color
     (a) 6   colorless   (b) 10   pink

Section 15.5 *Acid–Base Titrations*

27. $H_2SO_4(aq)$ + 2 NaOH(aq) $\rightarrow$ $Na_2SO_4(aq)$ + 2 $H_2O(l)$

28.15 ~~mL solution~~ × $\dfrac{0.100 \text{ mol NaOH}}{1000 \text{ ~~mL solution~~}}$ = 0.00282 mol NaOH

0.00282 ~~mol NaOH~~ × $\dfrac{1 \text{ mol } H_2SO_4}{2 \text{ ~~mol NaOH~~}}$ = 0.00141 mol H_2SO_4

$\dfrac{0.00141 \text{ mol } H_2SO_4}{10.0 \text{ ~~mL solution~~}}$ × $\dfrac{1000 \text{ ~~mL solution~~}}{1 \text{ L solution}}$ = $\dfrac{0.141 \text{ mol } H_2SO_4}{1 \text{ L solution}}$

Molarity of hydrochloric acid: 0.141 *M* H_2SO_4

29. $H_3PO_4(aq)$ + 3 NaOH(aq) $\rightarrow$ $Na_3PO_4(aq)$ + 3 $H_2O(l)$

34.45 ~~mL solution~~ × $\dfrac{0.210 \text{ mol NaOH}}{1000 \text{ ~~mL solution~~}}$ = 0.00723 mol NaOH

0.00723 ~~mol NaOH~~ × $\dfrac{1 \text{ mol } H_3PO_4}{3 \text{ ~~mol NaOH~~}}$ = 0.00241 mol H_3PO_4

$\dfrac{0.00241 \text{ mol } H_3PO_4}{50.0 \text{ ~~mL solution~~}}$ × $\dfrac{1000 \text{ ~~mL solution~~}}{1 \text{ L solution}}$ = $\dfrac{0.0482 \text{ mol } H_3PO_4}{1 \text{ L solution}}$

Molarity of phosphoric acid: 0.0482 *M* H_3PO_4

31. $H_2SO_3(aq)$ + 2 KOH(aq) $\rightarrow$ $K_2SO_3(aq)$ + 2 $H_2O(l)$

10.0 ~~mL solution~~ × $\dfrac{0.165 \text{ mol } H_2SO_3}{1000 \text{ ~~mL solution~~}}$ = 0.00165 mol H_2SO_3

0.00165 ~~mol H_2SO_3~~ × $\dfrac{2 \text{ mol KOH}}{1 \text{ ~~mol H_2SO_3~~}}$ = 0.00330 mol KOH

0.00330 ~~mol KOH~~ × $\dfrac{1000 \text{ mL solution}}{0.100 \text{ ~~mol KOH~~}}$ = 33.0 mL KOH

33.   (a)   MM of HCl $= 36.46$ g/mol

$$\frac{6.00 \text{ mol HCl}}{1000 \text{ mL solution}} \times \frac{36.46 \text{ g HCl}}{1 \text{ mol HCl}} \times \frac{1 \text{ mL solution}}{1.10 \text{ g solution}} \times 100\%$$

$$= 19.9\% \text{ HCl}$$

(b)   MM of $HC_2H_3O_2 = 60.06$ g/mol

$$\frac{1.00 \text{ mol } HC_2H_3O_2}{1000 \text{ mL solution}} \times \frac{60.06 \text{ g } HC_2H_3O_2}{1 \text{ mol } HC_2H_3O_2} \times \frac{1 \text{ mL solution}}{1.01 \text{ g solution}} \times 100\%$$

$$= 5.95\% \text{ } HC_2H_3O_2$$

(c)   MM of $HNO_3 = 63.02$ g/mol

$$\frac{0.500 \text{ mol } HNO_3}{1000 \text{ mL solution}} \times \frac{63.02 \text{ g } HNO_3}{1 \text{ mol } HNO_3} \times \frac{1 \text{ mL solution}}{1.01 \text{ g solution}} \times 100\%$$

$$= 3.12\% \text{ } HNO_3$$

(d)   MM of $H_2SO_4 = 98.09$ g/mol

$$\frac{3.00 \text{ mol } H_2SO_4}{1000 \text{ mL solution}} \times \frac{98.09 \text{ g } H_2SO_4}{1 \text{ mol } H_2SO_4} \times \frac{1 \text{ mL solution}}{1.18 \text{ g solution}} \times 100\%$$

$$= 24.9\% \text{ } H_2SO_4$$

Section 15.6 *Acid–Base Standardization*

35. $2 \text{ } HNO_3(aq) + Na_2CO_3(aq) \rightarrow 2 \text{ } NaNO_3(aq) + H_2O(l) + CO_2(g)$

MM of $Na_2CO_3 = 105.99$ g/mol

$$0.500 \text{ g } Na_2CO_3 \times \frac{1 \text{ mol } Na_2CO_3}{105.99 \text{ g } Na_2CO_3} \times \frac{2 \text{ mol } HNO_3}{1 \text{ mol } Na_2CO_3} = 0.00943 \text{ mol } HNO_3$$

$$\frac{0.00943 \text{ mol } HNO_3}{33.25 \text{ mL solution}} \times \frac{1000 \text{ mL solution}}{1 \text{ L solution}} = \frac{0.284 \text{ mol } HNO_3}{1 \text{ L solution}}$$

Molarity of nitric acid: $0.284 \text{ } M \text{ } HNO_3$

37. $2\,HCl(aq) + Na_2C_2O_4(aq) \rightarrow H_2C_2O_4(aq) + 2\,NaCl(aq)$

MM of $Na_2C_2O_4$ = 134.00 g/mol

$1.550\ \cancel{g\ Na_2C_2O_4} \times \dfrac{1\ \cancel{mol\ Na_2C_2O_4}}{134.00\ \cancel{g\ Na_2C_2O_4}} \times \dfrac{2\ mol\ HCl}{1\ \cancel{mol\ Na_2C_2O_4}} = 0.02313\ mol\ HCl$

$\dfrac{0.02313\ mol\ HCl}{20.95\ \cancel{mL\ solution}} \times \dfrac{1000\ \cancel{mL\ solution}}{1\ L\ solution} = \dfrac{1.104\ mol\ HCl}{1\ L\ solution}$

Molarity of hydrochloric acid: 1.104 M HCl

39. $H_2C_2O_4(aq) + 2\,LiOH(aq) \rightarrow Li_2C_2O_4(aq) + 2\,H_2O(l)$

MM of $H_2C_2O_4$ = 90.04 g/mol

$0.627\ \cancel{g\ H_2C_2O_4} \times \dfrac{1\ \cancel{mol\ H_2C_2O_4}}{90.04\ \cancel{g\ H_2C_2O_4}} \times \dfrac{2\ mol\ LiOH}{1\ \cancel{mol\ H_2C_2O_4}} = 0.0139\ mol\ LiOH$

$0.0139\ \cancel{mol\ LiOH} \times \dfrac{1000\ mL\ solution}{0.479\ \cancel{mol\ LiOH}} = 29.0\ mL\ solution$

41. $HAsc(aq) + NaOH(aq) \rightarrow NaAsc(aq) + H_2O(l)$

$30.95\ \cancel{mL\ solution} \times \dfrac{0.176\ \cancel{mol\ NaOH}}{1000\ \cancel{mL\ solution}} \times \dfrac{1\ mol\ HAsc}{1\ \cancel{mol\ NaOH}} = 0.00545\ mol\ HAsc$

MM of vitamin C = $\dfrac{0.959\ g\ HAsc}{0.00545\ mol\ HAsc} = 176\ g/mol$

43. 1 mole of alanine = 1 mole NaOH

$21.05\ \cancel{mL\ solution} \times \dfrac{0.145\ \cancel{mol\ NaOH}}{1000\ \cancel{mL\ solution}} \times \dfrac{1\ mol\ alanine}{1\ \cancel{mol\ NaOH}} = 0.00305\ mol\ alanine$

MM of alanine = $\dfrac{0.272\ g\ alanine}{0.00305\ mol\ alanine} = 89.2\ g/mol$

Section 15.7 *Ionization of Water*

45.   (a)   Simplified ionization equation: $H_2O(l) \rightarrow H^+(aq) + OH^-(aq)$
      (b)   Ionization constant expression: $K_W = [H^+] \, [OH^-]$
      (c)   Ionization constant for water at 25 °C: $K_W = 1.0 \times 10^{-14}$

47. $[OH^-] = \dfrac{1.0 \times 10^{-14}}{[H^+]}$

   (a) $[OH^-] = \dfrac{1.0 \times 10^{-14}}{0.025} = 4.0 \times 10^{-13}$

   (b) $[OH^-] = \dfrac{1.0 \times 10^{-14}}{0.000\ 017} = 5.9 \times 10^{-10}$

49. $[H^+] = \dfrac{1.0 \times 10^{-14}}{[OH^-]}$

   (a) $[H^+] = \dfrac{1.0 \times 10^{-14}}{0.0016} = 6.3 \times 10^{-12}$

   (b) $[H^+] = \dfrac{1.0 \times 10^{-14}}{0.000\ 29} = 3.4 \times 10^{-11}$

Section 15.8 *The pH Concept*

51. $pH = -\log [H^+]$

   (a) $pH = -\log 0.001 = -\log 10^{-3}$
         $pH = -(-3) = 3$

   (b) $pH = -\log 0.000\ 01 = -\log 10^{-5}$
         $pH = -(-5) = 5$

53. $[H^+] = 10^{-pH}$

   (a) $[H^+] = 10^{-6} = 0.000\ 001\ M\ (1 \times 10^{-6}\ M)$

   (b) $[H^+] = 10^{-8} = 0.000\ 000\ 01\ M\ (1 \times 10^{-8}\ M)$

Section 15.9 *pH Calculations*

55. (a) $pH = -\log 0.000\,000\,30 = -\log 3.0 \times 10^{-7}$
 $pH = -\log 3.0 - \log 10^{-7}$
 $pH = -0.48 - (-7) = 6.52$

 (b) $pH = -\log 0.000\,000\,016 = -\log 1.6 \times 10^{-8}$
 $pH = -\log 1.6 - \log 10^{-8}$
 $pH = -0.20 - (-8) = 7.80$

57. (a) $[H^+] = 10^{-1.80} = 10^{0.20} \times 10^{-2}$
 $[H^+] = 1.6 \times 10^{-2}\,M = 0.016\,M$

 (b) $[H^+] = 10^{-4.75} = 10^{0.25} \times 10^{-5}$
 $[H^+] = 1.8 \times 10^{-5}\,M = 0.000\,018\,M$

59. (a) $[OH^-] = 0.11\,M$

 $$[H^+] = \frac{1.0 \times 10^{-14}}{[OH^-]} = \frac{1.0 \times 10^{-14}}{0.11} = 9.1 \times 10^{-14}$$

 $pH = -\log (9.1 \times 10^{-14})$
 $pH = -\log 9.1 - \log 10^{-14}$
 $pH = -0.96 - (-14) = 13.04$

 (b) $[OH^-] = 0.000\,55\,M = 5.5 \times 10^{-4}\,M$

 $$[H^+] = \frac{1.0 \times 10^{-14}}{[OH^-]} = \frac{1.0 \times 10^{-14}}{5.5 \times 10^{-4}} = 1.8 \times 10^{-11}\,M$$

 $pH = -\log (1.8 \times 10^{-11})$
 $pH = -\log 1.8 - \log 10^{-11}$
 $pH = -0.26 - (-11) = 10.74$

61. (a) $pH = 0.90$
 $[H^+] = 10^{-0.90} = 10^{0.10} \times 10^{-1}$
 $[H^+] = 1.3 \times 10^{-1}\,M = 0.13\,M$

 $$[OH^-] = \frac{1.0 \times 10^{-14}}{[H^+]} = \frac{1.0 \times 10^{-14}}{0.13} = 7.7 \times 10^{-14}\,M$$

 (b) $pH = 1.62$
 $[H^+] = 10^{-1.62} = 10^{0.38} \times 10^{-2}$
 $[H^+] = 2.4 \times 10^{-2}\,M = 0.024\,M$

 $$[OH^-] = \frac{1.0 \times 10^{-14}}{[H^+]} = \frac{1.0 \times 10^{-14}}{0.024} = 4.2 \times 10^{-13}\,M$$

Section 15.10 *Strong and Weak Electrolytes*

63.

	Solution	Ionization
(a)	strong acids	highly ionized
(b)	strong bases	highly ionized
(c)	soluble ionic compounds	highly ionized

65.

	Solution	Electrolyte
(a)	$KOH(aq)$	strong
(b)	$NH_4OH(aq)$	weak
(c)	$Ca(OH)_2(aq)$	strong
(d)	$Mg(OH)_2(s)$	weak

67.

	Solution	Electrolyte
(a)	$ZnCO_3(s)$	weak
(b)	$Sr(NO_3)_2(aq)$	strong
(c)	$K_2SO_4(aq)$	strong
(d)	$PbI_2(s)$	weak

69.

	Solution	Electrolyte	Aqueous Solution
(a)	$HF(aq)$	weak	$HF(aq)$
(b)	$HBr(aq)$	strong	$H^+(aq)$ and $Br^-(aq)$
(c)	$HNO_3(aq)$	strong	$H^+(aq)$ and $NO_3^-(aq)$
(d)	$HNO_2(aq)$	weak	$HNO_2(aq)$

71.

	Solution	Electrolyte	Aqueous Solution
(a)	$AgF(aq)$	strong	$Ag^+(aq)$ and $F^-(aq)$
(b)	$AgI(s)$	weak	$AgI(s)$
(c)	$Hg_2Cl_2(s)$	weak	$Hg_2Cl_2(s)$
(d)	$NiCl_2(aq)$	strong	$Ni^{2+}(aq)$ and $2\,Cl^-(aq)$

Section 15.11 *Net Ionic Equations*

73.
(1) Complete and balance the nonionized equation.
(2) Convert the nonionized equation into a total ionic equation.
(3) Cancel spectator ions to obtain the net ionic equation.
(4) Check ($\sqrt{}$) each ion or atom on both sides of the equation.

75. (a) Nonionized equation:
$$HCl(aq) + KOH(aq) \rightarrow KCl(aq) + H_2O(l)$$

Total ionic equation:
$$H^+(aq) + Cl^-(aq) + K^+(aq) + OH^-(aq) \rightarrow K^+(aq) + Cl^-(aq) + H_2O(l)$$

Net ionic equation:
$$H^+(aq) + OH^-(aq) \rightarrow H_2O(l)$$

(b) Nonionized equation:
$$2\,HC_2H_3O_2(aq) + Ca(OH)_2(aq) \rightarrow Ca(C_2H_3O_2)_2(aq) + 2\,H_2O(l)$$

Total ionic equation:
$$2\,HC_2H_3O_2(aq) + Ca^{2+}(aq) + 2\,OH^-(aq) \rightarrow$$
$$Ca^{2+}(aq) + 2\,C_2H_3O_2^-(aq) + 2\,H_2O(l)$$

Net ionic equation:
$$HC_2H_3O_2(aq) + OH^-(aq) \rightarrow C_2H_3O_2^-(aq) + H_2O(l)$$

77. (a) Nonionized equation:
$$AgNO_3(aq) + KI(aq) \rightarrow AgI(s) + KNO_3(aq)$$

Total ionic equation:
$$Ag^+(aq) + NO_3^-(aq) + K^+(aq) + I^-(aq) \rightarrow K^+(aq) + NO_3^-(aq) + AgI(s)$$

Net ionic equation:
$$Ag^+(aq) + I^-(aq) \rightarrow AgI(s)$$

(b) Nonionized equation:
$$BaCl_2(aq) + K_2CrO_4(aq) \rightarrow BaCrO_4(s) + 2\,KCl(aq)$$

Total ionic equation:
$$Ba^{2+}(aq) + 2\,Cl^-(aq) + 2\,K^+(aq) + CrO_4^{2-}(aq) \rightarrow$$
$$BaCrO_4(s) + 2\,K^+(aq) + 2\,Cl^-(aq)$$

Net ionic equation:
$$Ba^{2+}(aq) + CrO_4^{2-}(aq) \rightarrow BaCrO_4(s)$$

General Exercises

79. The color of methyl red is yellow at pH 7, while phenolphthalein is colorless. Therefore, the color of the water with both indicators is *yellow*.

81. The color of methyl orange is red at pH 3.2 and yellow at pH 4.4. Therefore, the color of a solution at pH 3.8 is *orange* (red + yellow).

83. Notice that NaH_2PO_4 is a *proton acceptor* in the first reaction and a *proton donor* in the second reaction. Thus, NaH_2PO_4 is amphiprotic.

85. $42.10 \text{ mL solution} \times \dfrac{0.100 \text{ mol Ba(OH)}_2}{1000 \text{ mL solution}} \times \dfrac{1 \text{ mol cream of tartar}}{1 \text{ mol Ba(OH)}_2}$

$$= 0.00421 \text{ mol cream of tartar}$$

MM cream of tartar $= \dfrac{0.791 \text{ g cream of tartar}}{0.00421 \text{ mol cream of tartar}} = 188 \text{ g/mol}$

87. $HCl(aq) \rightarrow H^+(aq) + Cl^-(aq)$

87. $HCl(aq) \rightarrow H^+(aq) + Cl^-(aq)$

$0.50 \ M \ HCl \rightarrow 0.50 \ M \ H^+$

$pH = -\log [H^+]$
$pH = -\log 0.50$
$pH = 0.30$

89.   Nonionized equation:
$NH_4OH(aq) + HC_2H_3O_2(aq) \rightarrow NH_4C_2H_3O_2(aq) + H_2O(l)$

Total ionic equation:
$NH_4OH(aq) + HC_2H_3O_2(aq) \rightarrow NH_4^+(aq) + C_2H_3O_2^-(aq) + H_2O(l)$

Net ionic equation:
$NH_4OH(aq) + HC_2H_3O_2(aq) \rightarrow NH_4^+(aq) + C_2H_3O_2^-(aq) + H_2O(l)$

Chemical Equilibrium

Section 16.1 *Collision Theory*

1. Experimental Factors that Influence the Rate of Reaction
   (a) Yes, greater concentration of reactants increases the rate.
   (b) Yes, greater reaction temperature increases the rate.
   (c) Yes, by definition, a catalyst increases the reaction rate.
   (d) Occasionally, a few reactions are influenced by UV light.

3. Effective Collision Orientation

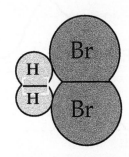

5.
Reaction Change	Effect on the Rate of Reaction
(a) increase the concentration of a reactant	The rate of reaction *increases* because more collisions occur.
(b) increase the temperature of the reaction	The rate of reaction *increases* because both collision frequency and collision energy decrease.
(c) add a metal catalyst	The rate of reaction *increases* because molecular collisions are more effective.

7. No, the catalyst speeds up the rate at which ammonia is produced, but does not increase the amount of ammonia produced.

9. $PCl_5(g)$ + *heat* $\rightleftarrows$ $PCl_3(g)$ + $Cl_2(g)$

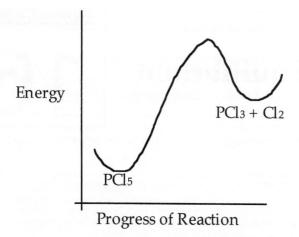

11. $H_2(g)$ + $I_2(g)$ + *heat* $\rightleftarrows$ $2\,HI(g)$

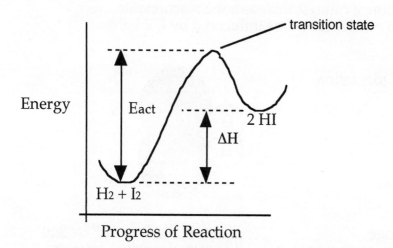

13.   A catalyst is a substance that allows a reaction to proceed faster by lowering the energy of activation (E_{act}).

15.   Since the reaction is exothermic, less energy is required for the reaction to proceed in the forward direction toward products. Thus, the E_{act} is greater in the reverse direction than in the forward direction.

Section 16.3 *The Chemical Equilibrium Concept*

17. (a) The rate of the forward reaction is measured by the rate of change in the decrease in concentration of reactant(s) in a given unit of time.

$$\text{rate of a forward reaction} = \frac{\text{decrease [reactant]}}{\text{unit time}}$$

(b) The rate of the forward reaction is measured by the rate of change in the increase in concentration of product(s) in a given unit of time.

$$\text{rate of a forward reaction} = \frac{\text{increase [product]}}{\text{unit time}}$$

19. (a) **False.** In the general equilibrium expression, the K_{eq} is dependent on temperature and its value varies at different temperatures.

(b) **True.** The amount of a solid does not affect a gaseous equilibrium, and thus is omitted from the K_{eq} expression. A substance participating in a different physical state (solid or liquid) is assigned a concentration of [1], and is incorporated into the K_{eq} value.

Section 16.4 *General Equilibrium Constant, K_{eq}*

21.

Equilibrium Reaction	Equilibrium Constant Expression
(a) $2\,A \rightleftharpoons C$	$K_{eq} = \dfrac{[C]}{[A]^2}$
(b) $A + 2\,B \rightleftharpoons 3\,C$	$K_{eq} = \dfrac{[C]^3}{[A]\,[B]^2}$
(c) $2\,A + 3\,B \rightleftharpoons 4\,C + D$	$K_{eq} = \dfrac{[C]^4\,[D]}{[A]^2\,[B]^3}$

23. Since the amount of any solid present has no effect on the equilibrium of a gaseous state reaction, any substance in the solid state will not appear in the equilibrium expression for a gaseous state reaction.

25.

Equilibrium Reaction	Equilibrium Constant Expression
(a) $H_2(g) + F_2(g) \rightleftharpoons 2\,HF(g)$	$K_{eq} = \dfrac{[HF]^2}{[H_2]\,[F_2]}$
(b) $4\,NH_3(g) + 7\,O_2(g) \rightleftharpoons 4\,NO_2(g) + 6\,H_2O(g)$	$K_{eq} = \dfrac{[NO_2]^4\,[H_2O]^6}{[NH_3]^4\,[O_2]^7}$
(c) $ZnCO_3(s) \rightleftharpoons ZnO(s) + CO_2(g)$	$K_{eq} = [CO_2]$

27. $2 SO_2(g) + O_2(g) \rightleftarrows 2 SO_3(g)$

$$[SO_2] = 1.75\,M \qquad [O_2] = 1.50\,M \qquad [SO_3] = 2.25\,M$$

$$K_{eq} = \frac{[SO_3]^2}{[SO_2]^2\,[O_2]} = \frac{(2.25)^2}{(1.75)^2\,(1.50)} = 1.10$$

Section 16.5 Equilibria Shifts for Gases

29. $\qquad N_2O_4(g) + heat \rightleftarrows 2 NO_2(g)$

(a) On hot days, the increase in heat shifts the equilibrium to the right producing more NO_2. Thus, the NO_2 concentration *increases*.

(b) On cool days, the decrease in heat shifts the equilibrium to the left producing less NO_2. Thus, the NO_2 concentration *decreases*.

31. $\qquad C(s) + CO_2(g) + heat \rightleftarrows 2 CO(g)$

(a)	increase $[CO_2]$	shifts right	(b)	decrease $[CO_2]$	shifts left
(c)	increase $[CO]$	shifts left	(d)	decrease $[CO]$	shifts right
(e)	increase temp.	shifts right	(f)	decrease temp.	shifts left
(g)	increase volume	shifts right	(h)	decrease volume	shifts left
(i)	add C powder	no shift	(j)	add Kr inert gas	no shift

33. $\qquad SO_2(g) + NO_2(g) \rightleftarrows SO_3(g) + NO(g) + heat$

(a)	increase $[SO_2]$	shifts right	(b)	decrease $[NO_2]$	shifts left
(c)	increase $[SO_3]$	shifts left	(d)	decrease $[NO]$	shifts right
(e)	increase temp.	shifts left	(f)	decrease temp.	shifts right
(g)	increase volume	no shift	(h)	decrease volume	no shift
(i)	add He inert gas	no shift	(j)	ultraviolet light	no shift

Section 16.6 Ionization Equilibrium Constant, K_i

35.   <u>Equilibrium Reaction</u>   <u>Equilibrium Constant Expression</u>

(a) $HCHO_2(aq) \rightleftarrows H^+(aq) + CHO_2^-(aq)$ $K_i = \dfrac{[H^+]\,[CHO_2^-]}{[HCHO_2]}$

(b) $H_2C_2O_4(aq) \rightleftarrows H^+(aq) + HC_2O_4^-(aq)$ $K_i = \dfrac{[H^+]\,[HC_2O_4^-]}{[H_2C_2O_4]}$

(c) $H_3C_6H_5O_7(aq) \rightleftarrows H^+(aq) + H_2C_6H_5O_7^-(aq)$ $K_i = \dfrac{[H^+]\,[H_2C_6H_5O_7^-]}{[H_3C_6H_5O_7]}$

37. $HNO_2(aq) \rightleftharpoons H^+(aq) + NO_2^-(aq)$

$$[H^+] = [NO_2^-] = 7.5 \times 10^{-3}\, M \qquad [HNO_2] = 0.125\, M$$

$$K_i = \frac{[H^+][NO_2^-]}{[HNO_2]} = \frac{(7.5 \times 10^{-3})(7.5 \times 10^{-3})}{(0.125)} = 4.5 \times 10^{-4}$$

39. $HF(aq) \rightleftharpoons H^+(aq) + F^-(aq)$

$$[H^+] = 10^{-pH} = 10^{-2.00} = 0.010\, M$$

$$[H^+] = [F^-] = 0.010\, M \qquad [HF] = 0.139\, M$$

$$K_i = \frac{[H^+][F^-]}{[HF]} = \frac{(0.010)(0.010)}{(0.139)} = 7.2 \times 10^{-4}$$

Section 16.7 *Equilibria Shifts for Weak Acids and Bases*

41. $\qquad HF(aq) \qquad \rightleftharpoons \qquad H^+(aq) + F^-(aq)$

(a)	increase [HF]	shift right	(b)	increase [H$^+$]	shift left
(c)	decrease [HF]	shift left	(d)	decrease [F$^-$]	shift right
(e)	add solid NaF	shift left	(f)	add gaseous HCl	shift left
(g)	add solid NaOH	shift right	(h)	increase pH	shift right

43. $\qquad HC_2H_3O_2(aq) \qquad \rightleftharpoons \qquad H^+(aq) + C_2H_3O_2^-(aq)$

(a)	increase [HC$_2$H$_3$O$_2$]	shift right	(b)	increase [H$^+$]	shift left
(c)	decrease [HC$_2$H$_3$O$_2$]	shift left	(d)	decrease [C$_2$H$_3$O$_2^-$]	shift right
(e)	add solid NaC$_2$H$_3$O$_2$	shift left	(f)	add solid NaCl	no shift
(g)	add NaOH solid	shift right	(h)	increase pH	shift right

Section 16.8 *Solubility Product Equilibrium Constant, K_{sp}*

45.
Reaction	Solubility Product Expression
(a) $AgI(s) \rightleftharpoons Ag^+(aq) + I^-(aq)$	$K_{sp} = [Ag^+][I^-]$
(b) $Ag_2CrO_4(s) \rightleftharpoons 2\,Ag^+(aq) + CrO_4^{2-}(aq)$	$K_{sp} = [Ag^+]^2[CrO_4^{2-}]$
(c) $Ag_3PO_4(s) \rightleftharpoons 3\,Ag^+(aq) + PO_4^{3-}(aq)$	$K_{sp} = [Ag^+]^3[PO_4^{3-}]$

47. $CoS(s) \rightleftarrows Co^{2+}(aq) + S^{2-}(aq)$

$[Co^{2+}] = [S^{2-}] = 7.7 \times 10^{-11}$

$K_{sp} = [Co^{2+}][S^{2-}] = (7.7 \times 10^{-11})(7.7 \times 10^{-11}) = 5.9 \times 10^{-21}$

49. $Zn_3(PO_4)_2(s) \rightleftarrows 3 Zn^{2+}(aq) + 2 PO_4^{3-}(aq)$

Since two moles of PO_4^{3-} are produced for every three moles of Zn^{2+},

then, $[PO_4^{3-}] = \frac{2}{3}[Zn^{2+}] = \frac{2}{3}(1.5 \times 10^{-7}) = 1.0 \times 10^{-7}$

$K_{sp} = [Zn^{2+}]^3 [PO_4^{3-}]^2 = (1.5 \times 10^{-7})^3 (1.0 \times 10^{-7})^2 = 3.4 \times 10^{-35}$

51. $CaCO_3(s) \rightleftarrows Ca^{2+}(aq) + CO_3^{2-}(aq)$

$[Ca^{2+}][CO_3^{2-}] = K_{sp} = 3.8 \times 10^{-9}$

Since one mole of CO_3^{2-} is produced for every mole of Ca^{2+},

then, $[Ca^{2+}]^2 = 3.8 \times 10^{-9}$

thus, $[Ca^{2+}] = 6.2 \times 10^{-5}$

$CaC_2O_4(s) \rightleftarrows Ca^{2+}(aq) + C_2O_4^{2-}(aq)$

$[Ca^{2+}][C_2O_4^{2-}] = K_{sp} = 2.3 \times 10^{-9}$

Since one mole of $C_2O_4^{2-}$ is produced for every mole of Ca^{2+},

then, $[Ca^{2+}]^2 = 2.3 \times 10^{-9}$

thus, $[Ca^{2+}] = 4.8 \times 10^{-5}$

The calcium ion concentration in a saturated solution of $CaCO_3$ is slightly greater than in a saturated solution of CaC_2O_4.

Section 16.9 Equilibria Shifts for Slightly Soluble Compounds

53. $Ca_3(PO_4)_2(s) \quad \rightleftarrows \quad 3\ Ca^{2+}(aq) + 2\ PO_4^{3-}(aq)$

(a) increase $[Ca^{2+}]$ shift left   (b) increase $[PO_4^{3-}]$ shift left
(c) decrease $[Ca^{2+}]$ shift right   (d) decrease $[PO_4^{3-}]$ shift right
(e) add solid $Ca(NO_3)_2$ shift left   (f) add solid KNO_3 no shift
(g) add solid $Ca_3(PO_4)_2$ no shift   (h) add H^+ shift right

55. $SrCO_3(s) \quad \rightleftarrows \quad Sr^{2+}(aq) + CO_3^{2-}(aq)$

(a) increase $[Sr^{2+}]$ shift left   (b) increase $[CO_3^{2-}]$ shift left
(c) decrease $[Sr^{2+}]$ shift right   (d) decrease $[CO_3^{2-}]$ shift right
(e) add solid $SrCO_3$ no shift   (f) add solid $Sr(NO_3)_2$ shift left
(g) add solid KNO_3 no shift   (h) decrease pH   shift right

General Exercises

57.   A swinging pendulum in a clock represents a *dynamic process* as it is in constant motion. The moving pendulum represents a *reversible process* as it swings back and forth in opposite directions.

59.   With regard to reaction rate, a reversible reaction at equilibrium has the following characteristic: *the rate of the forward reaction is equal to the rate of the reverse reaction.*

61. $N_2O_4(g) \quad \rightleftarrows \quad 2\ NO_2(g)$

$[N_2O_4] = 4.5 \times 10^{-5}\ M \qquad [NO_2] = 3.0 \times 10^{-3}\ M$

$$K_{eq} = \frac{[NO_2]^2}{[N_2O_4]} = \frac{(3.0 \times 10^{-3})^2}{(4.5 \times 10^{-5})} = 0.20$$

63. $H_2O(l) \quad \rightleftarrows \quad H^+(aq) + OH^-(aq)$

(a) increase $[H^+]$ shift left   (b) decrease $[OH^-]$ shift right
(c) increase pH   shift right   (d) decrease pH   shift left

65. $Fe(OH)_3(s) \rightleftarrows Fe^{3+}(aq) + 3\,OH^-(aq)$

Aqueous HCl ionizes to give $H^+(aq)$ and $Cl^-(aq)$. The $H^+(aq)$ can react with $OH^-(aq)$ to form H_2O. This decreases the $OH^-(aq)$ concentration which shifts the equilibrium to the right, thus allowing more dissociation of $Fe(OH)_3(s)$.

In summary, $Fe(OH)_3$ is more soluble in aqueous HCl than in water owing to the following reaction:

$$Fe(OH)_3(s) + 3\,H^+(aq) \rightleftarrows Fe^{3+}(aq) + 3\,H_2O(l)$$

67. $Ca(OH)_2(s) \rightleftarrows Ca^{2+}(aq) + 2\,OH^-(aq)$

$$pH = 12.35$$
$$[H^+] = 4.5 \times 10^{-13}\,M$$

$$[OH^-] = \frac{1.0 \times 10^{-14}}{[H^+]} = \frac{1.0 \times 10^{-14}}{4.5 \times 10^{-13}} = 2.2 \times 10^{-2}$$

Since two moles of OH^- are produced for every mole of Ca^{2+},

then, $[Ca^{2+}] = \frac{1}{2}[OH^-] = \frac{1}{2}(2.2 \times 10^{-2}) = 1.1 \times 10^{-2}$

Therefore,

$$K_{sp} = [Ca^{2+}][OH^-]^2 = (1.1 \times 10^{-2})(2.2 \times 10^{-2})^2 = 5.3 \times 10^{-6}$$

Oxidation and Reduction

CHAPTER 17

Section 17.1 *Oxidation Numbers*

1.

	Element	Ox No		Element	Ox No
(a)	Ag	0	(b)	Mg	0
(c)	S	0	(d)	F_2	0

(Note: All elements in the free state have an oxidation number of zero.)

3.

	Ion	Ox No		Ion	Ox No
(a)	Ag^+	+1	(b)	Mg^{2+}	+2
(c)	S^{2-}	–2	(d)	F^-	–1

5.

	Compound	Oxidation Number of Nitrogen
(a)	NH_3	ox no N + 3(ox no H) = 0 ox no N + 3(+1) = 0 ox no N + 3 = 0 ox no N = –3
(b)	N_2O_4	2(ox no N) + 4(ox no O) = 0 2(ox no N) + 4(–2) = 0 2(ox no N) + (–8) = 0 2(ox no N) = +8 ox no N = +4
(c)	Li_3N	3(ox no Li) + ox no N = 0 3(+1) + ox no N = 0 +3 + ox no N = 0 ox no N = –3

(d)　KNO_3　ox no K + ox no N + 3(ox no O) = 0
+1 + ox no N + 3(–2) = 0
+1 + ox no N + (–6) = 0
ox no N = +5

7.　
Ion	Oxidation Number of Sulfur

(a)　$SO_3{}^{2-}$　ox no S + 3(ox no O) = –2
ox no S + 3(–2) = –2
ox no S + (–6) = –2
ox no S = +4

(b)　$HSO_4{}^-$　ox no H + ox no S + 4(ox no O) = –1
(+1) + ox no S + 4(–2) = –1
(+1) + ox no S + (–8) = –1
ox no S = +6

(c)　HS^-　ox no H + ox no S = –1
(+1) + ox no S = –1
ox no S = –2

(d)　$S_2O_8{}^{2-}$　2(ox no S) + 8(ox no O) = –2
2(ox no S) + 8(–2) = –2
2(ox no S) + (–16) = –2
2(ox no S) = +14
ox no S = +7

Section 17.2 Oxidation–Reduction Reactions

9.　(a)　A redox process characterized by electron loss is termed *oxidation*.
　　(b)　A substance undergoing oxidation in a redox reaction is a *reducing agent*.

11.　(a)　Oxidized: Co　　　　　　Reduced: S
　　(b)　Oxidized: Cd　　　　　　Reduced: Cl_2

13.　(a)

$$CuO_{(s)} + H_{2(g)} \rightarrow Cu_{(s)} + H_2O_{(l)}$$

Oxidizing agent: CuO　　　　　Reducing agent: H_2

(b)

$$\overset{+2}{\underset{}{PbO_{(s)}}} + \overset{0}{CO_{(g)}} \rightarrow \overset{0}{Pb_{(s)}} + \overset{+4}{CO_{2(g)}}$$

Oxidizing agent: PbO   Reducing agent: CO

15. (a)

$$\overset{0}{Al_{(s)}} + \overset{+3}{Cr^{3+}_{(aq)}} \rightarrow \overset{+3}{Al^{3+}_{(aq)}} + \overset{0}{Cr_{(s)}}$$

Oxidized: Al   Reduced: Cr^{3+}

(b)

$$\overset{0}{F_{2(g)}} + 2\,\overset{-1}{Cl^-_{(aq)}} \rightarrow 2\,\overset{-1}{F^-_{(aq)}} + \overset{0}{Cl_{2(g)}}$$

Oxidized: Cl^- Reduced: F_2

17. (a)

$$\overset{+2}{Cr^{2+}_{(aq)}} + \overset{+1}{AgI_{(s)}} \rightarrow \overset{+3}{Cr^{3+}_{(aq)}} + \overset{0}{Ag_{(s)}} + I^-_{(aq)}$$

Oxidizing agent: AgI   Reducing agent: Cr^{2+}

(b)

$$\overset{+2}{Sn^{2+}_{(aq)}} + 2\,\overset{+2}{Hg^{2+}_{(aq)}} \rightarrow \overset{+4}{Sn^{4+}_{(aq)}} + \overset{+1}{Hg_2^{2+}_{(aq)}}$$

Oxidizing agent: Hg^{2+} Reducing agent: Sn^{2+}

Section 17.3 *Balancing Redox Equations: Oxidation Number Method*

19.   Yes, the total electron loss by oxidation must equal the total electron gain by the reduction process.

21.   (a)

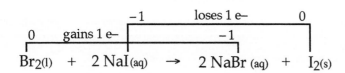

$$Br_{2(l)} + 2\,NaI_{(aq)} \rightarrow 2\,NaBr_{(aq)} + I_{2(s)}$$

(b)

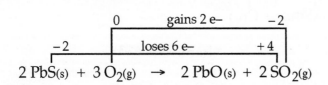

$$2\,PbS_{(s)} + 3\,O_{2(g)} \rightarrow 2\,PbO_{(s)} + 2\,SO_{2(g)}$$

23.   (a)

$$2\,MnO_4^-{}_{(aq)} + 10\,I^-{}_{(aq)} + 16\,H^+{}_{(aq)} \rightarrow 2\,Mn^{2+}{}_{(aq)} + 5\,I_{2(s)} + 8\,H_2O_{(l)}$$

(b)

$$Cu_{(s)} + 4\,H^+{}_{(aq)} + SO_4^{2-}{}_{(aq)} \rightarrow Cu^{2+}{}_{(aq)} + SO_{2(g)} + 2\,H_2O_{(l)}$$

Section 17.4 *Balancing Redox Equations: Half-Reaction Method*

25.   (a) $SO_2(g) + 2\,H_2O(l) \rightarrow SO_4^{2-}(aq) + 4\,H^+(aq) + 2\,e^-$
      (b) $2\,BrO_3^-(aq) + 12\,H^+(aq) + 10\,e^- \rightarrow Br_2(l) + 6\,H_2O(l)$

27.   (a) $ClO^-(aq) + H_2O(l) + 2\,e^- \rightarrow Cl^-(aq) + 2\,OH^-(aq)$
      (b) $MnO_4^-(aq) + 2\,H_2O(l) + 3\,e^- \rightarrow MnO_2(s) + 4\,OH^-(aq)$

29.   (a) Oxidation: $3\,(Zn \rightarrow Zn^{2+} + 2\,e^-)$
           Reduction: $2\,(NO_3^- + 4\,H^+ + 3\,e^- \rightarrow NO + 2\,H_2O)$

           $3\,Zn(s) + 2\,NO_3^-(aq) + 8\,H^+(aq) \rightarrow 3\,Zn^{2+}(aq) + 2\,NO(g) + 4\,H_2O(l)$

      (b) Oxidation: $2\,(4\,H_2O + Mn^{2+} \rightarrow MnO_4^- + 8\,H^+ + 5\,e^-)$
           Reduction: $5\,(BiO_3^- + 6\,H^+ + 2\,e^- \rightarrow Bi^{3+} + 3\,H_2O)$

           $2\,Mn^{2+}(aq) + 5\,BiO_3^-(aq) + 14\,H^+(aq) \rightarrow 2\,MnO_4^-(aq) + 5\,Bi^{3+}(aq) + 7\,H_2O(l)$

31.   (a)   Oxidation: $3 (S^{2-} \rightarrow S + 2 e^-)$
              Reduction: $2 (MnO_4^- + 2 H_2O + 3 e^- \rightarrow MnO_2 + 4 OH^-)$

              $3 S^{2-}(aq) + 2 MnO_4^-(aq) + 4 H_2O(l) \rightarrow 3 S(s) + 2 MnO_2(s) + 8 OH^-(aq)$

       (b)   Oxidation: $Cu \rightarrow Cu^{2+} + 2 e^-$
              Reduction: $ClO^- + H_2O + 2 e^- \rightarrow Cl^- + 2 OH^-$

              $Cu(s) + ClO^-(aq) + H_2O(l) \rightarrow Cu^{2+}(aq) + Cl^-(aq) + 2 OH^-(aq)$

33.   Oxidation: $Cl_2 + 2 H_2O \rightarrow 2 HOCl + 2 H^+ + 2 e^-$
              Reduction: $Cl_2 + 2 e^- \rightarrow 2 Cl^-$

              $Cl_2(g) + H_2O(l) \rightarrow Cl^-(aq) + HOCl(aq) + H^+(aq)$

35.   Oxidation: $Cl_2 + 8 OH^- \rightarrow 2 ClO_2^- + 4 H_2O + 6 e^-$
              Reduction: $3 (Cl_2 + 2 e^- \rightarrow 2 Cl^-)$

              $4 Cl_2(g) + 8 OH^-(aq) \rightarrow 2 ClO_2^-(aq) + 6 Cl^-(aq) + 4 H_2O(l)$

Section 17.5 *Predicting Spontaneous Redox Reactions*

37.
Substances	Greater Tendency to be Reduced
(a) $Pb^{2+}(aq)$ or $Zn^{2+}(aq)$	$Pb^{2+}(aq)$
(b) $Fe^{3+}(aq)$ or $Al^{3+}(aq)$	$Fe^{3+}(aq)$
(c) $Ag^+(aq)$ or $I_2(s)$	$Ag^+(aq)$
(d) $Cu^{2+}(aq)$ or $Br_2(l)$	$Br_2(l)$

39.
Substances	Stronger Oxidizing Agent
(a) $F_2(g)$ or $Cl_2(g)$	$F_2(g)$
(b) $Ag^+(aq)$ or $Br_2(l)$	$Br_2(l)$
(c) $Cu^{2+}(aq)$ or $H^+(aq)$	$Cu^{2+}(aq)$
(d) $Mg^{2+}(aq)$ or $Mn^{2+}(aq)$	$Mn^{2+}(aq)$

41.   (a)   nonspontaneous   (b)   spontaneous

43.   (a)   spontaneous   (b)   spontaneous

45. $A(s) + B^+(aq) \rightarrow A^+(aq) + B(s)$
       $B(s) + C^+(aq) \rightarrow B^+(aq) + C(s)$

       From the first reaction, we can conclude that A has a greater tendency to undergo oxidation than B. From the second reaction, we can conclude that B has a greater tendency to undergo oxidation than C. Therefore, the higher *reduction potentials* are as follows: C > B > A.

47.

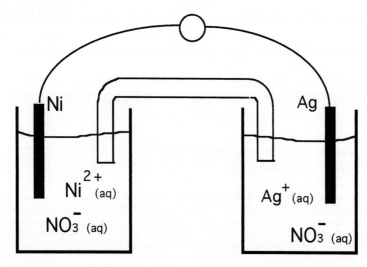

Anode: Ni   Cathode: Ag
$$Ni(s) + 2 AgNO_3(aq) \rightarrow 2 Ag(s) + Ni(NO_3)_2(aq)$$

49.

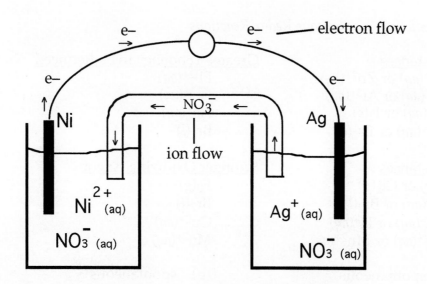

$$Ni(s) + 2 AgNO_3(aq) \rightarrow 2 Ag(s) + Ni(NO_3)_2(aq)$$

51. $Mn(s) + Fe^{2+}(aq) \rightarrow Mn^{2+}(aq) + Fe(aq)$

(a) oxidation half-cell reaction: $Mn(s) \rightarrow Mn^{2+}(aq) + 2 e^-$

(b) reduction half-cell reaction: $Fe^{2+}(aq) + 2 e^- \rightarrow Fe(aq)$

(c) anode: Mn electrode   cathode: Fe electrode

(d) direction of e^- flow: Mn anode $\rightarrow$ Fe cathode

(e) direction of SO_4^{2-} in the salt bridge: Fe half-cell $\rightarrow$ Mn half-cell

53. $Mg(s) + Pb(NO_3)_2(aq) \rightarrow Pb(s) + Mg(NO_3)_2(aq)$

    (a)   oxidation half-cell reaction: $Mg(s) \rightarrow Mg^{2+}(aq) + 2\ e^-$

    (b)   reduction half-cell reaction: $Pb^{2+}(aq) + 2\ e^- \rightarrow Pb(s)$

    (c)   anode: Mg electrode   cathode: Pb electrode

    (d)   direction of e^- flow: Mg anode $\rightarrow$ Pb cathode

    (e)   direction of NO_3^- in the salt bridge: Pb half-cell $\rightarrow$ Mg half-cell

Section 17.7 *Electrolytic Cells*

55. $Ni(s) + Fe^{2+}(aq) \rightarrow Fe(s) + Ni^{2+}(aq)$

    (a)   oxidation half-cell reaction: $Ni(s) \rightarrow Ni^{2+}(aq) + 2\ e^-$

    (b)   reduction half-cell reaction: $Fe^{2+}(aq) + 2\ e^- \rightarrow Fe(s)$

    (c)   anode: Ni electrode   cathode: Fe electrode

    (d)   direction of e^- flow: Ni anode $\rightarrow$ Fe cathode

    (e)   direction of SO_4^{2-} in the salt bridge: Fe half-cell $\rightarrow$ Ni half-cell

57. $Pb(s) + Zn^{2+}(aq) \rightarrow Zn(s) + Pb^{2+}(aq)$

    (a)   oxidation half-cell reaction: $Pb(s) \rightarrow Pb^{2+}(aq) + 2\ e^-$

    (b)   reduction half-cell reaction: $Zn^{2+}(aq) + 2\ e^- \rightarrow Zn(s)$

    (c)   anode: Pb electrode   cathode: Zn electrode

    (d)   direction of e^- flow: Pb anode $\rightarrow$ Zn cathode

    (e)   direction of NO_3^- in the salt bridge: Zn half-cell $\rightarrow$ Pb half-cell

59. $Cl_2(l) + 2\,NaF(aq) \rightarrow 2\,NaCl(aq) + F_2(g)$

    (a)   oxidation half-cell reaction: $2\,F^-(aq) \rightarrow F_2(g) + 2\ e^-$

    (b)   reduction half-cell reaction: $Cl_2(g) + 2\ e^- \rightarrow 2\,Cl^-(aq)$

    (c)   anode: $Pt(F_2)$ electrode   cathode: $Pt(Cl_2)$ electrode

    (d)   direction of e^- flow: $Pt(F_2)$ anode $\rightarrow$ $Pt(Cl_2)$ cathode

    (e)   direction of Na^+ in the salt bridge: $Pt(F_2)$ half-cell $\rightarrow$ $Pt(Cl_2)$ half-cell

General Exercises

61. <u>Calcium Thiosulfate</u>   <u>Oxidation Number of Sulfur</u>
      CaS_2O_3 (ox no Ca) + 2(ox no S) + 3(ox no O) = 0
                                    (+2) + 2(ox no S) + 3(−2) = 0
                                    (+2) + 2(ox no S) + (−6) = 0
                                    2(ox no S) = −2 +6 = +4
                                    ox no S = +2

63. Net ionic equation:
     $Co(s) + Hg^{2+}(aq) \rightarrow Co^{2+}(aq) + Hg(l)$

65. Net ionic equation:
     $Zn(s) + 2\,H^+(aq) \rightarrow Zn^{2+}(aq) + H_2(g)$

67. Net ionic equation:
     $2\,Li(s) + 2\,H_2O(l) \rightarrow 2\,Li^+(aq) + 2\,OH^-(aq) + H_2(g)$

69. Net ionic equation:
     $3\,Cu(s) + 8\,H^+(aq) + 2\,NO_3^-(aq) \rightarrow 3\,Cu^{2+}(aq) + 2\,NO(g) + 4\,H_2O(l)$

Nuclear Chemistry

Section 18.1 *Natural Radioactivity*

1. An alpha particle (α) is a helium nucleus, which is deflected toward the negative electrode as it passes between electrically charged plates.

3. A beta particle (β) is identical to an electron, which is deflected toward the positive electrode as it passes between electrically charged plates.

5. A gamma ray (γ) is light energy, which is *not* deflected as it passes between electrically charged plates

Section 18.2 *Nuclear Equations*

7.
Radiation	Approximate Mass
(a) alpha particle (α)	~4 amu
(c) beta particle (β^-)	~0 amu
(b) gamma ray (γ)	0 amu
(d) positron (β^+)	~0 amu
(e) neutron (n^o)	~1 amu
(f) proton (p^+)	~1 amu

9. (a) $^{160}_{74}\text{W} \rightarrow ^{156}_{72}\text{Hf} + ^{4}_{2}\text{He}$

    (b) $^{32}_{15}\text{P} \rightarrow ^{32}_{16}\text{S} + ^{0}_{-1}\text{e}$

    (c) $^{55}_{27}\text{Co} \rightarrow ^{55}_{26}\text{Fe} + ^{0}_{+1}\text{e}$

    (d) $^{44}_{22}\text{Ti} + ^{0}_{-1}\text{e} \rightarrow ^{44}_{21}\text{Sc}$

11.   (a) $^{221}_{88}X \rightarrow {}^{217}_{86}Rn + {}^{4}_{2}He$ $^{221}_{88}X = {}^{221}_{88}Ra$

         (b) $^{43}_{19}X \rightarrow {}^{43}_{20}Ca + {}^{0}_{-1}e$ $^{43}_{19}X = {}^{43}_{19}K$

         (c) $^{73}_{36}X \rightarrow {}^{73}_{35}Br + {}^{0}_{+1}e$ $^{73}_{36}X = {}^{73}_{36}Kr$

         (d) $^{133}_{56}X + {}^{0}_{-1}e \rightarrow {}^{133}_{55}Cs$ $^{133}_{56}X = {}^{133}_{56}Ba$

Section 18.3 *Radioactive Decay Series*

13. $^{238}_{92}U \rightarrow {}^{234}_{90}Th + {}^{4}_{2}He$

        $^{234}_{90}Th \rightarrow {}^{234}_{91}Pa + {}^{0}_{-1}e$ The resulting nuclide is: $^{234}_{91}$Pa.

15.   Let X = the parent nuclide:

        $^{210}_{84}X \rightarrow {}^{206}_{82}Pb + {}^{4}_{2}He$ The parent nuclide is: $^{210}_{84}$Po.

17.   <u>Decay Series for Thorium-232</u>   <u>Answers</u>

$^{232}_{90}Th \xrightarrow{\alpha} {}^{228}_{88}Ra \xrightarrow{\beta} {}^{228}_{89}Ac$

                                   $\downarrow \beta$

$^{220}_{86}Rn \xleftarrow{\alpha} {}^{224}_{88}Ra \xleftarrow{\alpha} {}^{228}_{90}Th$

$\alpha \downarrow$

$^{216}_{84}Po \xrightarrow{\alpha} {}^{212}_{82}Pb \xrightarrow{\beta} {}^{212}_{83}Bi$

                                   $\downarrow \beta$

$^{208}_{82}Pb \xleftarrow{\alpha} {}^{212}_{84}Po$

Answers:

(a) α (b) β

(c) β (d) α

(e) α (f) α

(g) α (h) β

(i) β (j) α

19. $^{241}_{94}Pu \rightarrow {}^{241}_{95}Am + {}^{0}_{-1}e$

        $^{241}_{95}Am \rightarrow {}^{237}_{93}Np + {}^{4}_{2}He$

        $^{237}_{93}Np \rightarrow {}^{233}_{91}Pa + {}^{4}_{2}He$

Section 18.4 *Radioactive Half-Life*

21. $100\% \times \frac{1}{2} \times \frac{1}{2} = 25\%$

    25% of the U-238 radionuclides remain after two half-lives.

23. Pu-239: $\quad 15\ t_{1/2} \times \dfrac{24,400 \text{ years}}{1\ t_{1/2}} = 366,000 \text{ years}$

25.

Activity of C-14	Elapsed Time
100%	$0\ t_{1/2}$
50%	$1\ t_{1/2}$
25%	$2\ t_{1/2}$
12.5%	$3\ t_{1/2}$
6.25%	$4\ t_{1/2}$

$$4\ t_{1/2} \times \dfrac{5730 \text{ years}}{1\ t_{1/2}} \approx 22,900 \text{ years old}$$

27. $60 \text{ hours} \times \dfrac{1\ t_{1/2}}{15 \text{ hours}} = 4\ t_{1/2}$

    $80 \text{ mg Na-24} \times \frac{1}{2} \times \frac{1}{2} \times \frac{1}{2} \times \frac{1}{2} = 5 \text{ mg Na-24}$

29.

Activity of Ir-192	Elapsed Time
560 dpm	$0\ t_{1/2}$
280 dpm	$1\ t_{1/2}$
140 dpm	$2\ t_{1/2}$
70 dpm	$3\ t_{1/2}$
35 dpm	$4\ t_{1/2}$

$$4\ t_{1/2} \times \dfrac{74 \text{ days}}{1\ t_{1/2}} = 296 \text{ days}$$

31.

Activity of Co-60	Elapsed Time
1200 dpm	$0\ t_{1/2}$
600 dpm	$1\ t_{1/2}$
300 dpm	$2\ t_{1/2}$
150 dpm	$3\ t_{1/2}$
75 dpm	$4\ t_{1/2}$

$$\text{half-life} = \dfrac{21.2 \text{ years}}{4\ t_{1/2}} = 5.30 \text{ years/half-life}$$

Section 18.5 *Applications of Radionuclides*

33. Uranium–lead dating can be used to estimate geological events occurring billions of years ago.

35. The γ-emitting radionuclide cobalt-60 can be used to sterilize male insects for agricultural pest control.

37. Plutonium-238 releases high-energy gamma rays that are used to power heart pacemakers.

39. The β-emitting radionuclide iodine-131 can be used to measure the activity of the thyroid gland.

41. The γ-emitting radionuclide technetium-99 can be used to diagnose and locate brain tumors.

Section 18.6 *Induced Radioactivity*

43. $^{81}_{35}\text{Br} + ^{0}_{0}\gamma \rightarrow ^{80}_{35}\text{Br} + ^{1}_{0}\text{n}$ The radionuclide is: $^{80}_{35}\text{Br}$.

45. $^{238}_{92}\text{U} + ^{2}_{1}\text{H} \rightarrow ^{238}_{94}\text{Pu} + 2\,^{1}_{0}\text{n} + ^{0}_{-1}\text{e}$ The radionuclide is: $^{238}_{94}\text{Pu}$.

47. $^{130}_{52}\text{Te} + ^{2}_{1}\text{H} \rightarrow ^{131}_{53}\text{I} + ^{1}_{0}\text{n}$ The target nuclide is: $^{130}_{52}\text{Te}$.

49. $^{10}_{5}\text{B} + ^{4}_{2}\text{He} \rightarrow ^{14}_{7}\text{N} + ^{0}_{0}\gamma$ The projectile particle is: $^{4}_{2}\text{He}$.

51. $^{243}_{95}\text{Am} + ^{22}_{10}\text{Ne} \rightarrow ^{260}_{105}\text{Db} + 5\,^{1}_{0}\text{n}$ The Russians created: $^{260}_{105}\text{Db}$.

53. $^{54}_{24}\text{Cr} + ^{207}_{82}\text{Pb} \rightarrow ^{260}_{106}\text{Sg} + ^{1}_{0}\text{n}$ The Russians created: $^{260}_{106}\text{Sg}$.

Section 18.7 Nuclear Fission

55. $$^{1}_{0}n \xrightarrow{\text{1st}} 2\,^{1}_{0}n \xrightarrow{\text{2nd}} 4\,^{1}_{0}n \xrightarrow{\text{3rd}} 8\,^{1}_{0}n$$

After the third step, 8 neutrons are produced.

57. Uranium-235 can fission in many different ways to give various products. The number of neutrons released from a single fission is usually 2 or 3, but the *average* number of neutrons is 2.4.

59. $$^{235}_{92}U + ^{1}_{0}n \rightarrow ^{142}_{56}Ba + ^{91}_{36}Kr + 3\,^{1}_{0}n \qquad \text{The number of neutrons is 3.}$$

61. $$^{239}_{94}Pu + ^{1}_{0}n \rightarrow ^{137}_{55}Cs + ^{100}_{39}Y + 3\,^{1}_{0}n \qquad \text{The number of neutrons is 3.}$$

63. $$^{233}_{92}U + ^{1}_{0}n \rightarrow ^{143}_{56}Ba + ^{88}_{36}Kr + 3\,^{1}_{0}n \qquad \text{The fissionable nuclide is: } ^{233}_{92}U.$$

Section 18.8 Nuclear Fusion

65. $$^{2}_{1}H + ^{2}_{1}H \rightarrow ^{3}_{1}H + ^{0}_{+1}e + ^{1}_{0}X \qquad \text{The particle } X \text{ is: } ^{1}_{0}n.$$

67. $$^{3}_{2}He + ^{1}_{1}X \rightarrow ^{4}_{2}He + ^{0}_{+1}e \qquad \text{The particle } X \text{ is: } ^{1}_{1}H.$$

69. $$^{2}_{1}X + ^{2}_{1}X \rightarrow ^{4}_{2}He + ^{0}_{0}\gamma \qquad \text{The unknown nuclide is: } ^{2}_{1}H.$$

General Exercises

71. $$^{241}_{94}Pu \rightarrow ^{4}_{2}He + ^{237}_{92}X \qquad \text{The daughter product is: } ^{237}_{92}U.$$

73. $$^{137}_{53}X \rightarrow ^{136}_{54}Xe + ^{0}_{-1}e + ^{1}_{0}n \qquad \text{The radionuclide } X \text{ is: } ^{137}_{53}I.$$

75. Since an activity of ~7.7 dpm per gram indicates that one half-life has expired, the age of Crater Lake is ~6000 years because $1\,t_{1/2} = 5730$ years.

77. $^{241}_{95}X \rightarrow {}^{237}_{93}Np + {}^{4}_{2}He$ The radionuclide is: $^{241}_{95}Am$.

79. $^{238}_{92}U + 15\,{}^{1}_{0}n \rightarrow {}^{253}_{99}Es + 7\,{}^{0}_{-1}e$ The number of neutrons is 15.

81. $^{253}_{99}Es + {}^{4}_{2}He \rightarrow {}^{1}_{0}n + {}^{256}_{101}X$ The nuclide X is: $^{256}_{101}Md$.

83. $^{2}_{1}H + {}^{3}_{1}H \rightarrow {}^{4}_{2}He + {}^{1}_{0}n + energy$ The emitted particle is: $^{1}_{0}n$.

Organic Chemistry

Section 19.1 *Hydrocarbons*

1.   The three types of fossil fuels are petroleum, coal, and natural gas.

3.   The alkane class of compounds consists of saturated hydrocarbons.

5.   The alkene and alkyne classes of compounds are unsaturated.

Section 19.2 *Alkanes*

7.   Molecular Formula   Class of Compound

   (a) $C_{10}H_{24}$ The molecular formula of the alkanes
                                  is C_nH_{2n+2}. Since $C_{10}H_{24}$ does *not* fit
                                  the general formula, it is *not* an alkane.

   (b) $C_{10}H_{22}$ The molecular formula of the alkanes
                                  is C_nH_{2n+2}. Since $C_{10}H_{22}$ does fit the
                                  general formula, it is an alkane.

9.   Structural Formula   IUPAC Name
   (a) $CH_3-CH_2-CH_3$ propane
   (b) $CH_3-CH_2-CH_2-CH_2-CH_3$ pentane
   (c) $CH_3-CH_2-CH_2-CH_2-CH_2-CH_2-CH_3$ heptane
   (d) $CH_3-CH_2-CH_2-CH_2-CH_2-CH_2-CH_2-CH_2-CH_3$ nonane

11. <u>Five Isomers of Hexane</u> (C_6H_{14})

$CH_3 - CH_2 - CH_2 - CH_2 - CH_2 - CH_3$

$$CH_3 - \overset{\overset{\displaystyle CH_3}{|}}{CH} - CH_2 - CH_2 - CH_3$$

$$CH_3 - CH_2 - \overset{\overset{\displaystyle CH_3}{|}}{CH} - CH_2 - CH_3$$

$$CH_3 - \overset{\overset{\displaystyle CH_3}{|}}{CH} - \overset{\overset{\displaystyle CH_3}{|}}{CH} - CH_3$$

$$CH_3 - \overset{\overset{\displaystyle CH_3}{|}}{\underset{\underset{\displaystyle CH_3}{|}}{C}} - CH_2 - CH_3$$

13. <u>Two Isomers of Chloropropane</u> (C_3H_7Cl)

$CH_3–CH_2–CH_2–Cl$ $CH_3–CH(Cl)–CH_3$

15.

<u>Alkyl Group</u>	<u>IUPAC Name</u>
(a) methyl	$CH_3–$
(b) ethyl	$CH_3CH_2–$

17. (a) 2-methylpentane   (b) 4-ethyl-2-methylhexane
     (c) 2,4,4-trimethylheptane   (d) 3,4,5-trimethyloctane

19. (a) $CH_4 + 2O_2 \rightarrow CO_2 + 2H_2O$
     (b) $C_3H_8 + 5O_2 \rightarrow 3CO_2 + 4H_2O$
     (c) $2C_6H_{14} + 19O_2 \rightarrow 12CO_2 + 14H_2O$
     (d) $2C_8H_{18} + 25O_2 \rightarrow 16CO_2 + 18H_2O$

Section 19.3 *Alkenes and Alkynes*

21.

	<u>Molecular Formula</u>	<u>Class of Compound</u>
(a)	$C_{10}H_{22}$	The molecular formula of the alkenes is C_nH_{2n}. Since $C_{10}H_{22}$ does *not* fit the general formula, it is *not* an alkene.
(b)	$C_{10}H_{20}$	The molecular formula of the alkenes is C_nH_{2n}. Since $C_{10}H_{20}$ does fit the general formula, it is an alkene.

23.

	Structural Formula	IUPAC Name
(a)	$CH_2=CH-CH_2-CH_3$	1-butene
(b)	$CH_2=CH-CH_2-CH_2-CH_3$	1-pentene
(c)	$CH_3-CH=CH-CH_2-CH_2-CH_3$	2-hexene
(d)	$CH_3-CH_2-CH_2-CH=CH-CH_2-CH_2-CH_3$	4-octene

25.

	Molecular Formula	Class of Compound
(a)	$C_{10}H_{20}$	The molecular formula of the alkynes is C_nH_{2n-2}. Since $C_{10}H_{20}$ does *not* fit the general formula, it is *not* an alkyne.
(b)	$C_{10}H_{18}$	The molecular formula of the alkynes is C_nH_{2n-2}. Since $C_{10}H_{18}$ does fit the general formula, it is an alkyne.

27.

	Structural Formula	IUPAC Name
(a)	$CH_3-C \equiv C-CH_2-CH_3$	2-pentyne
(b)	$CH \equiv C-CH_2-CH_2-CH_3$	1-pentyne
(c)	$CH_3-CH_2-C \equiv C-CH_2-CH_3$	3-hexyne
(d)	$CH_3-C \equiv C-CH_2-CH_2-CH_2-CH_3$	2-heptyne

29. <u>Two Isomers of Straight-Chain Pentene</u> (C_5H_{10})

$$CH_2 = CH-CH_2-CH_2-CH_3 \qquad CH_3-CH = CH-CH_2-CH_3$$

1-pentene   2-pentene

31. (a)

$$CH_3$$
$$|$$
$$CH_3-CH=CH-CH-CH_3$$

4-methyl-2-pentene

(b)

$$CH_3 \qquad CH_3$$
$$| \qquad\quad |$$
$$CH_3-CH-CH_2-C=CH-CH_3$$

3,5-dimethyl-2-hexene

33. (a)

$$\begin{array}{c} \quad\quad\quad CH_3 \\ \quad\quad\quad | \\ CH_3{-}C \equiv C{-}C{-}CH_2{-}CH_2{-}CH_3 \\ \quad\quad\quad | \\ \quad\quad\quad CH_3 \end{array}$$

4,4-dimethyl-2-heptyne

(b)

$$\begin{array}{c} \quad\quad\quad\quad CH_3 \\ \quad\quad\quad\quad | \\ CH_3{-}CH_2{-}CH_2{-}CH{-}CH{-}CH{-}C \equiv CH \\ \quad\quad | \quad\quad\quad | \\ \quad\quad CH_3 \quad\quad CH_3 \end{array}$$

3,4,5-trimethyl-1-octyne

35. (a) $CH_2 = CH_2 + 3\,O_2 \rightarrow 2\,CO_2 + 2\,H_2O$

(b) $CH_3{-}CH = CH_2 + H_2 \rightarrow CH_3{-}CH_2{-}CH_3$

(c) $CH_3{-}CH = CH{-}CH_3 + Br_2 \rightarrow CH_3{-}CHBr{-}CHBr{-}CH_3$

Section 19.4 *Arenes*

37. <u>Three Isomers of Xylene</u> $[C_6H_4(CH_3)_2]$

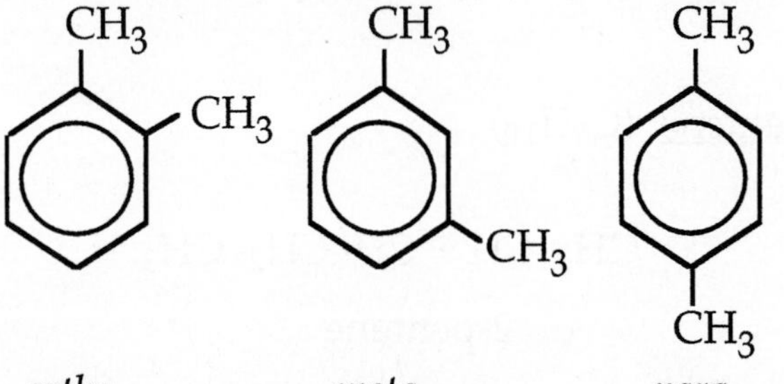

ortho *meta* *para*

39. $C_6H_6 + O_2 \xrightarrow{\text{spark}} CO_2 + H_2O$ (carbon dioxide and water)

Section 19.5 *Hydrocarbon Derivatives*

41.

General Formula	Class of Compound
(a) R—O—R′	ether
(b) R—X	organic halide
(c) Ar—OH	phenol
(d) $R{-}NH_2$	amine

43. | Chemical Formula | Class of Compound |

(a) $CH_3–NH_2$ amine

(b) $CH_3CH_2–F$ organic halide

(c)
$$CH_3-\overset{\overset{\textstyle O}{\|}}{C}-O-CH_3$$
ester

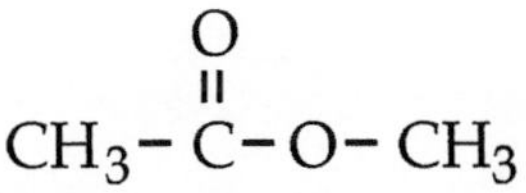

(d)
$$H-\overset{\overset{\textstyle O}{\|}}{C}-OH$$
carboxylic acid

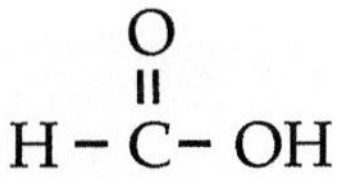

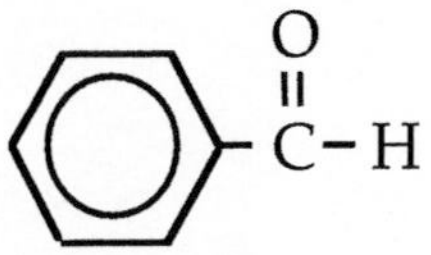

(e)
aldehyde

(f) CH_3CH_2-O-
ether

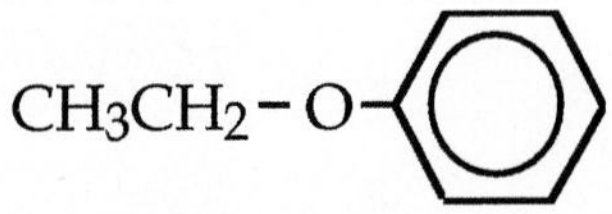

(g)
ketone

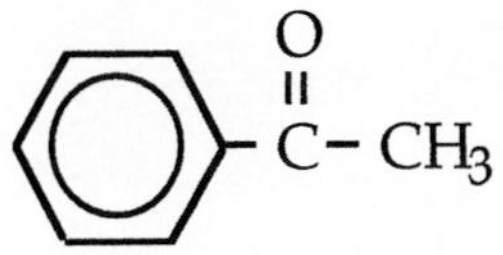

(h)
amide

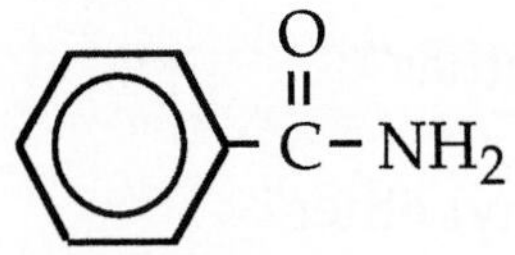

Section 19.6 *Organic Halides*

45. | Organic Halide | IUPAC Name | Common Name |
|---|---|---|
| (a) $CH_3–I$ | iodomethane | "methyl iodide" |
| (b) $CH_3–CH_2–Br$ | bromoethane | "ethyl bromide" |
| (c) $CH_3–CH_2–CH_2–F$ | 1-fluoropropane | "propyl fluoride" |
| (d) $(CH_3)_2–CH–Cl$ | 2-chloropropane | "isopropyl chloride" |

47.   1,1,2-trichloroethylene $CCl_2=CHCl$

49.   Solvent   Solute Solubility
     (a)   water   organic halides insoluble in water
     (b)   hydrocarbons   organic halides soluble in hydrocarbons

Section 19.7 *Alcohols, Phenols, and Ethers*

51.   Alcohol   IUPAC Name
     (a) $CH_3-CH_2-CH_2-CH_2-OH$ 1-butanol
     (b) $CH_3-CH_2-CH(OH)-CH_3$ 2-butanol
     (c) $CH_2(OH)-CH_2-CH_2-CH_3$ 1-butanol
     (d) $CH_3-CH(OH)-CH_2-CH_3$ 2-butanol

53.   Phenol   IUPAC Name

     (a)   —OH   phenol

     (b) CH_3——OH   *para*-methylphenol

55.   Ether   Common Name

     (a) CH_3-O-CH_3 "dimethyl ether"

     (b) $CH_3-O-CH_2CH_3$ "methyl ethyl ether"

     (c) $CH_3CH_2CH_2-O-CH_2CH_2CH_3$ "dipropyl ether"

     (d)   —$O-CH_2CH_3$ "phenyl ethyl ether"

57. Although "ethyl alcohol" and "dimethyl ether" have the same molecular mass, the alcohol has a higher boiling point. Alcohols have strong intermolecular attraction due to hydrogen bonding, while ethers have weaker dipole attraction.

59.   Alcohol (C_2H_6O)   Ether (C_2H_6O)

       CH_3-CH_2-OH CH_3-O-CH_3

Section 19.8 *Amines*

61.   Amine   IUPAC Name   Common Name
       (a) $CH_3-CH_2-NH_2$ amino ethane   "ethylamine"
       (b) $CH_3-CH_2-CH_2-NH_2$ 1-amino propane   "propylamine"

63. Although "ethylamine" and propane have about the same molecular mass, "ethylamine" has a higher boiling point. Amines have a strong intermolecular attraction due to hydrogen bonding, while alkanes have a weak intermolecular attraction due to dispersion forces.

Section 19.9 *Aldehydes and Ketones*

65.   Aldehyde   IUPAC Name   Common Name

       (a) $$\begin{array}{c} O \\ \parallel \\ H-C-H \end{array}$$ methanal   "formaldehyde"

       (b) $$\begin{array}{c} O \\ \parallel \\ CH_3-C-H \end{array}$$ ethanal   "acetaldehyde"

       (c) $$\begin{array}{c} O \\ \parallel \\ CH_3CH_2-C-H \end{array}$$ propanal   "propionaldehyde"

       (d) $$\begin{array}{c} O \\ \parallel \\ CH_3CH_2CH_2-C-H \end{array}$$ butanal   "butyraldehyde"

67. | Ketone | IUPAC Name | Common Name |
|---|---|---|

(a)
$$O$$
$$\|$$
$$CH_3- C- CH_3$$
2-propanone   "dimethyl ketone"

(b)
$$O$$
$$\|$$
$$CH_3- C- CH_2CH_3$$
2-butanone   "methyl ethyl ketone"

(c)
$$O$$
$$\|$$
$$CH_3- C- CH_2CH_2CH_3$$
2-pentanone   "methyl propyl ketone"

(d)
$$O$$
$$\|$$
$$CH_3CH_2- C- CH_2CH_3$$
3-pentanone   "diethyl ketone"

69. (a) Aldehydes have *higher boiling points* than hydrocarbons with similar molecular mass. Aldehydes demonstrate a strong intermolecular attraction due to dipole forces, while hydrocarbons are attracted by weaker dispersion forces.

(b) Ketones have *higher boiling points* than hydrocarbons with similar molecular mass. Ketones demonstrate a strong intermolecular attraction due to dipole forces, while hydrocarbons are attracted by weaker dispersion forces.

71. Aldehyde (C_3H_6O)   Ketone (C_3H_6O)

$$O$$
$$\|$$
$$CH_3CH_2-C-H$$

$$O$$
$$\|$$
$$CH_3-C-CH_3$$

Section 19.10 *Carboxylic Acids, Esters, and Amides*

73. | Carboxylic Acid | IUPAC Name | Common Name |
|---|---|---|

(a)
$$O$$
$$\|$$
$$H- C- OH$$
methanoic acid   "formic acid"

(b)
$$O$$
$$\|$$
$$CH_3- C- OH$$
ethanoic acid   "acetic acid"

(c)
$$O$$
$$\|$$
$$CH_3CH_2- C- OH$$
propanoic acid   "propionic acid"

(d)
$$O$$
$$\|$$
$$CH_3CH_2CH_2- C- OH$$
butanoic acid   "butyric acid"

75. | Ester | IUPAC Name | Common Name |

(a)

$$\overset{\displaystyle O}{\overset{\displaystyle \|}{H-C}}-O-CH_2CH_3$$

ethyl methanoate   "ethyl formate"

(b)

$$CH_3-\overset{\displaystyle O}{\overset{\displaystyle \|}{C}}-O-CH_3$$

methyl ethanoate   "methyl acetate"

(c)

$$CH_3CH_2-\overset{\displaystyle O}{\overset{\displaystyle \|}{C}}-O-CH_2CH_3$$

ethyl propanoate   "ethyl propionate"

(d)

$$H-\overset{\displaystyle O}{\overset{\displaystyle \|}{C}}-O-\bigcirc$$

phenyl methanoate   "phenyl formate"

77. | Amide | IUPAC Name | Common Name |

(a)

$$H-\overset{\displaystyle O}{\overset{\displaystyle \|}{C}}-NH_2$$

methanamide   "formamide"

(b)

$$\bigcirc-\overset{\displaystyle O}{\overset{\displaystyle \|}{C}}-NH_2$$

benzamide   —

79. (a) Carboxylic acids have an –OH group, and thus can *hydrogen bond*.
    (b) Esters do not have an –OH or –NH$_2$ group, and cannot *hydrogen bond*.
    (c) Amides have an –NH$_2$ group, and thus can *hydrogen bond*.

81. Although "propionic acid" and "methyl acetate" have the same molecular mass, "propionic acid" has a *higher boiling point*. Acids have a strong intermolecular attraction due to hydrogen bonding, while esters are attracted only by weaker dipole forces.

83. (a)

$$CH_3-\overset{\overset{\displaystyle O}{\|}}{C}-OH \;+\; CH_3-OH \;\;\xrightarrow{\;H_2SO_4\;}\;\; CH_3-\overset{\overset{\displaystyle O}{\|}}{C}-O-CH_3 \;+\; H_2O$$

(b)

$$C_6H_5-\overset{\overset{\displaystyle O}{\|}}{C}-OH \;+\; NH_3 \;\;\xrightarrow{\;\Delta\;}\;\; C_6H_5-\overset{\overset{\displaystyle O}{\|}}{C}-NH_2 \;+\; H_2O$$

85. The ester produced from the reaction of phenol and ethanoic acid is phenyl ethanoate ("phenyl acetate").

87. The amide produced from the reaction of ammonia and methanoic acid is methanamide ("formamide").

General Exercises

89.

Compound	Classification
(a) eicosane	saturated (–ane suffix)
(b) dodecene	unsaturated (–ene suffix)

91.

Compounds	Higher Boiling Point
(a) CH_3-O-CH_3 or CH_3CH_2-OH	CH_3CH_2-OH
(b) $CH_3CH_2CH_2-NH_2$ or $CH_3CH_2CH_2-F$	$CH_3CH_2CH_2-NH_2$

In each of the above pairs, the compound with the higher boiling point has an –OH or –NH_2 group, which can hydrogen bond. Hydrogen bonds increase the effective molecular mass of the compound, thus requiring more energy (that is, higher temperature) to boil.

93. Acetylsalicylic acid has an –ic acid suffix; thus, the functional group present is a carboxylic acid.

95. Androsterone has an –one suffix; thus, the functional group present is a ketone.

97. Chloral hydrate has a chlor– prefix and has an –al suffix; thus, the functional groups present are an organic halide (Cl) and an aldehyde.

Biochemistry

CHAPTER

20

Section 20.1 *Biological Compounds*

1. (a) A *protein* is a polymer consisting of many amino acids.
    (b) A *nucleic acid* is a polymer consisting of many nucleotides, which each contain a sugar, a base, and phosphoric acid.

3. (a) The repeating units in a protein are joined by *peptide linkages*.
    (b) The repeating units in a nucleic acid are joined by *phosphate linkages*.

5. | Biological Compound | Organic Functional Group |
    |---|---|
    | (a) protein | carboxylic acid, amine, amide |
    | (b) carbohydrate | aldehyde or ketone, alcohol |
    | (c) lipid | carboxylic acid, alcohol, ester |
    | (d) nucleic acid | aldehyde, alcohol, amine |

Section 20.2 *Proteins*

7. A covalent amide bond is responsible for linking two amino acids together in the primary structure of a protein.

9. The primary structure of a protein refers to the amino acid sequence. The secondary structure of a protein refers to the shape of the chain, for example, an α-helix or a pleated sheet.

11. The dipeptide formed from two molecules of alanine is named alanylalanine (Ala–Ala):

$$H_2N-CH-\overset{\overset{\displaystyle O}{\|}}{C}-NH-CH-\overset{\overset{\displaystyle O}{\|}}{C}-OH$$
$$\underset{\displaystyle CH_3}{|} \qquad\qquad \underset{\displaystyle CH_3}{|}$$

13. The dipeptide formed from a molecule of serine and a molecule of tyrosine is named seryltyrosine (Ser–Tyr):

$$H_2N-CH-\underset{\underset{O}{\|}}{C}-NH-CH-\underset{\underset{O}{\|}}{C}-OH$$

with side chains:
CH₂–OH (serine) and CH₂–(phenyl-OH) (tyrosine)

15.

Tripeptide	Amino Acid Sequence
1	Gly–His–Trp
2	Gly–Trp–His
3	His–Gly–Trp
4	His–Trp–Gly
5	Trp–Gly–His
6	Trp–His–Gly

17. The three amino acids in the segment of protein chain are *glutamic acid, lysine,* and *serine.*

19. Two of the amino acids in oxytocin differ from those in vasopressin. In oxytocin, *isoleucine* replaces phenylalanine, and *leucine* replaces arginine.

Section 20.3 *Enzymes*

21. In the lock-and-key model, the lock represents the *substrate (S)* molecule.

23. The term for a location on an enzyme that conforms to the shape of the substrate molecule is called an *active site.*

25. In Step 1 of enzyme catalysis, the substrate molecule binds to the active site on the enzyme.

27. An enzyme is likely to be inactive under experimental conditions where the pH or temperature is too high or too low.

Section 20.4 *Carbohydrates*

29.   An aldose sugar has an *aldehyde* functional group as well as one or more *alcohol* groups.

31.   A hexose sugar molecule contains six carbon atoms.

33.   A monosaccharide is a simple sugar molecule such as glucose, whereas a disaccharide has two simple sugar molecules joined by a glycoside linkage.

35.   See textbook Figure 20.9.

37.   See textbook Figure 20.11.

39.
$$C_{12}H_{22}O_{11} \quad + \quad H_2O \quad \overset{H^+}{\rightarrow} \quad C_6H_{12}O_6 \quad + \quad C_6H_{12}O_6$$

    maltose   water   glucose   glucose

41.   A glucose sugar molecule is the repeating monomer in glycogen, a small polysaccharide.

Section 20.5 *Lipids*

43.

Constituents	Lipid
(a) saturated fatty acids	more typical of a fat
(b) unsaturated fatty acids	more typical of an oil
(c) glycerol	fat and oil
(d) ester linkages	fat and oil

45.   The structural formula for the triglyceride of stearic acid is:

$$
\begin{array}{l}
\qquad\qquad\quad \overset{\displaystyle O}{\overset{\|}{}} \\
CH_2\!-\!O\!-\!C\text{-}(CH_2)_{16}\!-\!CH_3 \\
\;| \qquad\qquad \overset{\displaystyle O}{\overset{\|}{}} \\
CH\,-\,O\!-\!C\text{-}(CH_2)_{16}\!-\!CH_3 \\
\;| \qquad\qquad \overset{\displaystyle O}{\overset{\|}{}} \\
CH_2\!-\!O\!-\!C\text{-}(CH_2)_{16}\!-\!CH_3
\end{array}
$$

Stearic acid: $CH_3\text{-}(CH_2)_{16}\text{-}COOH$

47.

Name of Fatty Acid	Structure of Fatty Acid
(a) palmitic acid	$HOOC-(CH_2)_{14}-CH_3$
(b) lauric acid	$HOOC-(CH_2)_{10}-CH_3$
(c) myristic acid	$HOOC-(CH_2)_{12}-CH_3$

49. The saponification of carnauba wax with aqueous NaOH produces a fatty acid and long-chain alcohol.

$$CH_3-(CH_2)_{24}-\overset{\overset{\displaystyle O}{||}}{C}-O-(CH_2)_{29}-CH_3 \;+\; NaOH \;\rightarrow$$

$$CH_3-(CH_2)_{24}-\overset{\overset{\displaystyle O}{||}}{C}-O^- \; Na^+ \;+\; CH_3-(CH_2)_{29}-OH$$

51. The water-insoluble vitamins (A, D, E, K) are considered lipids.

Section 20.6 *Nucleic Acids*

53. The three general components of a DNA nucleotide are deoxyribose sugar, a base, and phosphoric acid.

55. The four bases in DNA are adenine (A), cytosine (C), guanine (G), and thymine (T).

57. The structures of deoxyribose (in DNA) and ribose (in RNA) are identical except that deoxyribose is missing a hydroxyl (–OH) group.

59. A DNA molecule has two polymer strands of nucleotides in the shape of a double helix.

61. There are two hydrogen bonds between adenine and thymine (A=T).

63. During DNA replication, an adenine base (A) on the template strand codes for an thymine base (T) base on the complementary strand.

65. During RNA transcription, an adenine base (A) on the template DNA strand codes for a uracil base (U) on the growing RNA strand.

General Exercises

67. A protein that provides strength, such as that in muscle tissue, has a long and extended shape.

69. When a polysaccharide undergoes hydrolysis in aqueous acid, glycoside linkages are cleaved, releasing the monosaccharide sugars in the polymer.

71. Since the compound is a polyhydroxy aldehyde polymer, the main component in an insect skeleton is *carbohydrate*.

73. There are six possible codons having an adenine base (A), a cytosine base (C), and a guanine base (G): ACG, AGC, CAG, CGA, GAC, and GCA.

75. The first step in *replication* is for a DNA molecule to unwind by breaking the hydrogen bonds (A=T and C≡G) that hold the double helix together.

   In the second step, each strand of DNA acts as a template to synthesize a complementary strand of DNA. That is, a nucleotide in the template strand that contains the base adenine (A) codes for thymine (T) on the growing strand; cytosine (C) codes for guanine (G); guanine (G) codes for cytosine (C); and thymine (T) codes for adenine (A).

   At the conclusion, each pair of template strands and complementary strands are joined by hydrogen bonds (A=T and C≡G) and the two resulting structures are exact duplicates of the original DNA molecule. Thus, two identical molecules of DNA are produced from one molecule of DNA.

77. DNA imprints genetic information onto RNA, which in turn translates the genetic code into instructions for protein synthesis. First, a molecule of DNA in the cell nucleus synthesizes a molecule of RNA. Next, the RNA molecule moves out of the nucleus into the surrounding cytoplasm, where protein synthesis actually occurs.

   The genetic code involves trinucleotides in the RNA molecule, which are referred to as codons. For example, the CAU codon represents three consecutive nucleotides that contain the bases cytosine, adenine, and uracil, respectively. The CAU codon is specific for the amino acid *histidine*, and is responsible for adding histidine to a growing protein chain. Other examples include the GAG codon, which specifies *glutamine*; and the UUU codon, which specifies *phenylalanine*.

Flashcards

for

Common Monoatomic and Polyatomic

Cations and Anions

$$Mg^{2+}$$

$$Ba^{2+}$$

$$Ca^{2+}$$

$$Sr^{2+}$$

$$Na^+$$

$$K^+$$

$$H^+$$

$$Li^+$$

barium ion	magnesium ion
strontium ion	calcium ion
potassium ion	sodium ion
lithium ion	hydrogen ion

$$Al^{3+}$$

$$Mn^{2+}$$

$$Ni^{2+}$$

$$Cr^{3+}$$

$$Ag^{+}$$

$$Cd^{2+}$$

$$Zn^{2+}$$

manganese(II) ion

aluminum ion

chromium(III) ion

nickel(II) ion

cadmium ion

silver ion

zinc ion

Cu^+	Cu^{2+}
Fe^{2+}	Fe^{3+}
Pb^{2+}	Pb^{4+}
Sn^{2+}	Sn^{4+}

copper(II) ion
cupric ion

copper(I) ion
cuprous ion

iron(III) ion
ferric ion

iron(II) ion
ferrous ion

lead(IV) ion
plumbic ion

lead(II) ion
plumbous ion

tin(IV) ion
stannic ion

tin(II) ion
stannous ion

$$Co^{2+}$$

$$Co^{3+}$$

$$Hg_2^{2+}$$

$$Hg^{2+}$$

$$NH_4^+$$

$$H_3O^+$$

$$CN^-$$

$$OH^-$$

chloride ion	fluoride ion
iodide ion	bromide ion
sulfide ion	oxide ion
phosphide ion	nitride ion

ClO_4^-

ClO^-

ClO_3^-

ClO_2^-

NO_3^-

NO_2^-

SO_4^{2-}

SO_3^{2-}

hydrogen carbonate ion	**carbonate ion**
phosphate ion	**hydrogen sulfate ion**
permanganate ion	**acetate ion**
dichromate ion	**chromate ion**